Biomechanics of Soft and Hard Tissues

Biomechanics of Soft and Hard Tissues

Guest Editors

Cecilia Surace
Alice Berardo

MDPI

Basel • Beijing • Wuhan • Barcelona • Belgrade • Novi Sad • Cluj • Manchester

Guest Editors

Cecilia Surace
Department of Structural, Geotechnical and Building Engineering
Politecnico di Torino
Turin
Italy

Alice Berardo
Department of Industrial Engineering
University of Padova
Padova
Italy

Editorial Office
MDPI AG
Grosspeteranlage 5
4052 Basel, Switzerland

This is a reprint of the Special Issue, published open access by the journal *Applied Sciences* (ISSN 2076-3417), freely accessible at: https://www.mdpi.com/journal/applsci/special_issues/68006EG7G6.

For citation purposes, cite each article independently as indicated on the article page online and as indicated below:

Lastname, A.A.; Lastname, B.B. Article Title. *Journal Name* **Year**, *Volume Number*, Page Range.

ISBN 978-3-7258-5191-1 (Hbk)
ISBN 978-3-7258-5192-8 (PDF)
https://doi.org/10.3390/books978-3-7258-5192-8

Cover image courtesy of Alice Berardo

Contents

About the Editors

Cecilia Surace

Cecilia Surace is Associate Professor of Structural Mechanics at Politecnico di Torino, where she leads the BIOMAST Lab (BIO-MAterials and STructures). Her research at BIOMAST specialises in applying structural mechanics principles to biomechanics. The lab focuses on the mechanical behaviour of biological tissues and medical devices, combining structural modelling, numerical simulation, and experimental testing to develop innovative solutions. Under her guidance, the lab has developed novel tissue-repair devices for tendons and the abdominal wall, offering suture-free alternatives to conventional techniques.

Beyond biomechanics, Dr Surace's research spans a broad range of structural engineering topics, including inverse methods, damage identification, and structural health monitoring. She actively collaborates on interdisciplinary projects and has a strong commitment to technology transfer.

Alice Berardo

Alice Berardo is a research fellow in Industrial Bioengineering at the Department of Industrial Engineering at the University of Padova (Italy) and a member of the interdepartmental Centre for Mechanics of Biological Materials (CMBM). She holds a PhD in Solid Mechanics from the University of Trento (Italy), where she focused on anisotropic modelling of friction phenomena and mechanical characterization of bioinspired materials.

Her current research focuses on both experimental and computational biomechanics, with particular attention to the mechanical behaviour of biological materials in healthy and pathological conditions. She works extensively on soft tissues—such as gastric, fascial, and adipose tissue—as well as hard tissues like bone.

Her interdisciplinary approach combines mechanical testing, material modelling, and numerical simulations, contributing to the understanding of tissue mechanics and the support of biomedical innovation.

applied sciences

MDPI

Editorial

Special Issue "Biomechanics of Soft and Hard Tissues"

Alice Berardo [1,*] and Cecilia Surace [2]

[1] Department of Industrial Engineering, University of Padova, 35131 Padova, Italy
[2] BIO-MAterials and STructures Laboratory (BIOMAST Lab), Department of Structural, Geotechnical and Building Engineering, Politecnico di Torino, 10129 Turin, Italy; cecilia.surace@polito.it
* Correspondence: alice.berardo@unipd.it

1. Introduction

The field of biomechanics has witnessed a profound evolution over the past decades, increasingly serving as a vital bridge between engineering principles and biological sciences. As biomedical challenges become more complex and patient-specific, the integration of mechanical analysis and modelling with medical research has become indispensable. This Special Issue, "Biomechanics of Soft and Hard Tissues", hosted by *Applied Sciences*, reflects this interdisciplinary landscape, bringing together novel studies that span experimental testing, constitutive modelling, and advanced fabrication techniques such as additive manufacturing.

Soft and hard biological tissues, while differing significantly in their structure and mechanical behaviour, both play critical roles in maintaining physiological function and structural integrity. From the compliant nature of ligaments and skin to the load-bearing stiffness of bone and dental tissues, these materials respond to mechanical stimuli in highly nonlinear and often hierarchical ways [1–3]. Understanding and replicating these responses remain central goals in biomedical engineering, with implications ranging from injury prevention and prosthetic design to tissue engineering and regenerative medicine [4–6].

In recent years, a discernible trend has emerged in the biomechanics community: the drive to integrate sophisticated experimental methodologies with advanced computational modelling [7,8]. This synergy enables not only more accurate simulations of physiological behaviour but also the development of predictive tools for personalised medicine. At the same time, advances in materials science and fabrication technologies have unlocked new pathways for replicating or replacing biological tissues with unprecedented fidelity [9,10]. Additive manufacturing, in particular, has become a game-changing tool, allowing for the production of anatomically accurate, mechanically tuneable, and biocompatible structures that were previously impossible to fabricate.

This Special Issue captures the state-of-the-art in this rapidly evolving field. The nine original contributions selected for publication reflect the diversity and depth of current research in biomechanics. They explore innovative strategies for mechanical characterisation of tissues and biomaterials, propose new constitutive frameworks for describing nonlinear and anisotropic behaviours, and demonstrate the practical integration of emerging manufacturing methods. The studies also collectively highlight the importance of rigorous validation—whether through in vitro testing, in silico simulations, or comparisons with clinical data.

One of the central themes that emerges from this collection is the necessity of interdisciplinary collaboration. No single field—whether mechanical engineering, materials science, biology, or clinical medicine—can alone address the complexity of the mechanical behaviour of living tissues. The contributions in this Special Issue exemplify collaborative efforts that draw on complementary expertise: combining detailed microstructural analysis with continuum modelling, or merging experimental mechanics with biomedical imaging and design.

This Special Issue also emphasises the growing importance of multiscale and multiphysics approaches. Similarly, the mechanical behaviour of tissues is often governed not only by macroscopic forces but also by microscale architecture, cellular activity, and even molecular interactions. Addressing these phenomena requires new modelling paradigms and experimental protocols, which the articles in this Special Issue begin to explore.

Moreover, the inclusion of additive manufacturing across multiple contributions signals an important direction for the future of biomedical engineering. 3D printing techniques enable not only the fabrication of patient-specific implants and scaffolds but also the realisation of mechanical gradients and anisotropic properties that are characteristic of native tissues. Importantly, the development of bioinspired and biomimetic materials through such technologies holds promise for improving long-term clinical outcomes, particularly when integrated with tailored mechanical and biological cues.

As we consider the broader implications of this Special Issue, several key insights emerge. First, mechanical testing remains a cornerstone of biomechanics yet it must continuously evolve to capture the complex behaviour of biological tissues under physiological and pathological conditions. This requires both methodological innovation and greater standardisation across laboratories and disciplines. Second, constitutive modelling is entering a new phase, where phenomenological laws are increasingly complemented by structurally informed, micro–macro approaches. These developments are enabling more realistic simulations of tissue deformation, growth, and remodelling, with applications ranging from surgical planning to implant optimisation. Third, fabrication techniques are becoming an active design variable, not just a final step in prototyping. The ability to control mechanical properties through layer-by-layer deposition or material blending introduces a new design space in tissue engineering and biomechanics, one that integrates mechanics from the very beginning of the process.

Despite the remarkable advances represented in this Special Issue, challenges remain. Experimental reproducibility, model validation, and clinical translation all require continued attention. Likewise, ethical and regulatory considerations must be addressed as personalised and mechanically tailored solutions move closer to clinical deployment. These challenges, however, should not deter innovation; rather, they should encourage closer cooperation between researchers, clinicians, and industry stakeholders.

2. Overview of Contributions

The contributions collected in this reprint converge on a common objective: to deepen our understanding of the biomechanical behaviour of soft and hard tissues through a combination of experimental methods, computational modelling, and engineering innovation. Although diverse in focus, the articles form a cohesive narrative that highlights three key research threads: mechanical characterisation of biological tissues, modelling and simulation for physiological understanding and clinical insight, and the development of biomimetic or engineered solutions for tissue repair and regeneration.

A first group of papers is centred on the mechanical characterisation of biological tissues, offering valuable data and insights into the structural and functional complexity of human anatomy. Berardo et al. [contribution 4] conduct a unique case study analysing multiple soft tissues from the same human male donor within the lower urinary tract, providing comparative mechanical data essential for pelvic modelling. Similarly, Peña-Trabalon et al. [contribution 3] assess surrogate models for meniscal root repair by comparing porcine and aged human tissues, contributing critical guidance for translational orthopaedic biomechanics. Sahinis and Kellis [contribution 8] use ultrasound imaging to explore the relationship between hamstring muscle and aponeurosis thickness, advancing our understanding of muscle–tendon interactions relevant to both sports science and rehabilitation.

The second thematic thread involves computational modelling and physiological simulation, where biomechanical methods are used to investigate complex biological functions and pathologies. Rădulescu et al. [contribution 1] propose a novel radiographic–stochastic model (URCOTEBS) to describe the complementary states of health and disease in the D-organ and middle-ear mucosa, combining imaging data with probabilistic modelling to support diagnostic applications. Minniti et al. [contribution 9] simulate blood flow in the aorta and right coronary artery through computational fluid dynamics, providing insights into vascular behaviour under both physiological and pathological conditions. Yamin et al. [contribution 7] perform a biomechanical stress analysis on the ankle joint under varying inclinations, contributing to the understanding of joint loading during posture and movement.

A third group of contributions focuses on biomimetic design and regenerative approaches, integrating biomechanics with fabrication technologies. Dolino et al. [contribution 2] present a synthetic 3D-printed cartilage scaffold for the human knee, combining microstructural design, finite element modelling, and mechanical testing to develop a biomimetic solution for joint repair. Belgio et al. [contribution 6] explore 3D bioprinting to recreate the extracellular microenvironment necessary for retinal regeneration, demonstrating how mechanical cues can be engineered to support cell function. Finally, Burgio et al. [contribution 5] offer a comprehensive review of mechanical stapling devices used in soft tissue repair, assessing commercially available solutions and providing a foundation for future improvements in surgical device design.

3. Concluding Remarks

The studies published in this Special Issue reflect the wide scope and interdisciplinary depth of current biomechanical research. Together, these studies embody the interdisciplinary nature of modern biomechanics. They demonstrate how experimental, computational, and engineering approaches can be successfully integrated to investigate biological complexity, support clinical decision-making, and design therapeutic solutions. Whether through characterising native tissue properties, simulating physiological processes, or engineering replacements, each contribution reinforces the importance of biomechanics at the interface of medicine, biology, and technology.

In summary, they illustrate the essential role of biomechanics in addressing challenges at the interface of engineering, biology, and medicine. We hope that this collection will serve as a valuable reference and inspire new collaborations in the rapidly evolving field of tissue biomechanics.

Author Contributions: Conceptualisation, A.B. and C.S.; methodology, A.B. and C.S.; investigation, A.B.; data curation, C.S.; writing—original draft preparation, A.B.; writing—review and editing, C.S. All authors have read and agreed to the published version of the manuscript.

Funding: This research received no external funding.

Acknowledgments: We extend our sincere gratitude to the contributing authors and peer reviewers for their dedication and scholarly excellence. We also thank the *Applied Sciences* Editorial Office for their support throughout the publication process.

Conflicts of Interest: The authors declare no conflicts of interest.

List of Contributions:

1. Rădulescu, M.; Mocanu, A.; Neagu, A.; Schipor, M.; Mocanu, H. Defining the UR-COTEBS System as a Unilateral Radiographic–Stochastic Model for the Complementary States (Health/Disease) of the D-Organ and Middle-Ear Mucosa. *Appl. Sci.* **2023**, *13*, 12861.
2. Dolino, G.; Coato, D.; Forni, R.; Boretti, G.; Ciliberti, F.; Gargiulo, P. Designing a Synthetic 3D-Printed Knee Cartilage: FEA Model, Micro-Structure and Mechanical Characteristics. *Appl. Sci.* **2024**, *14*, 331.
3. Peña-Trabalon, A.; Perez-Blanca, A.; Moreno-Vegas, S.; Estebanez-Campos, M.; Prado-Novoa, M. Assessment of Surrogate Models for Research on Resistance and Deformation of Repairs of the Human Meniscal Roots: Porcine or Older Human Models? *Appl. Sci.* **2024**, *14*, 670.
4. Berardo, A.; Mascolini, M.; Fontanella, C.; Contran, M.; Todesco, M.; Porzionato, A.; Macchi, V.; De Caro, R.; Boscolo-Berto, R.; Carniel, E. Mechanical Characterization of the Male Lower Urinary Tract: Comparison among Soft Tissues from the Same Human Case Study. *Appl. Sci.* **2024**, *14*, 1357.
5. Burgio, V.; Bei, J.; Rodriguez Reinoso, M.; Civera, M.; Grimaldo Ruiz, O.; Surace, C.; Pugno, N. Mechanical Stapling Devices for Soft Tissue Repair: A Review of Commercially Available Linear, Linear Cutting, and Circular Staplers. *Appl. Sci.* **2024**, *14*, 2486.
6. Belgio, B.; Mantero, S.; Iervolino, F.; Potere, F.; Levi, M.; Boschetti, F. Bioprinting of a Biomimetic Microenvironment for a Retinal Regenerative Approach. *Appl. Sci.* **2024**, *14*, 6980.
7. Yamin, N.; Basaruddin, K.; Ijaz, M.; Mat Som, M.; Shahrol Aman, M.; Takemura, H. Stress Analysis on the Ankle Joint during Incline and Decline Standing. *Appl. Sci.* **2024**, *14*, 8168.
8. Sahinis, C.; Kellis, E. Does a Greater Hamstring Muscle Thickness Mean a Greater Aponeurosis Thickness? *Appl. Sci.* **2025**, *15*, 1520.
9. Minniti, M.; Gramigna, V.; Palumbo, A.; Fragomeni, G. Application of Mathematical Models for Blood Flow in Aorta and Right Coronary Artery. *Appl. Sci.* **2025**, *15*, 5748.

References

1. Singh, G.; Chanda, A. Mechanical properties of whole-body soft human tissues: A review. *Biomed. Mater.* **2021**, *16*, 062004. [CrossRef] [PubMed]
2. Holzapfel, G.A. *Nonlinear Solid Mechanics: A Continuum Approach for Engineering*; Wiley: Hoboken, NJ, USA, 2000.
3. Bonaldi, L.; Berardo, A.; Pirri, C.; Stecco, C.; Carniel, E.L.; Fontanella, C.G. Mechanical Characterization of Human Fascia Lata: Uniaxial Tensile Tests from Fresh-Frozen Cadaver Samples and Constitutive Modelling. *Bioengineering* **2023**, *10*, 226. [CrossRef] [PubMed]
4. Fung, Y.C. *Biomechanics: Mechanical Properties of Living Tissues*; Springer: Berlin/Heidelberg, Germany, 1993.
5. Humphrey, J.D.; Delange, S.L. *An Introduction to Biomechanics: Solids and Fluids, Analysis and Design*; Springer: Berlin/Heidelberg, Germany, 2004.
6. Holzapfel, G.A.; Ogden, R.W. *Biomechanics of Soft Tissue in Cardiovascular Systems*; Springer: Berlin/Heidelberg, Germany, 2003.
7. Pandy, M.G. Computer modeling and simulation of human movement. *Annu. Rev. Biomed. Eng.* **2001**, *3*, 245–273. [CrossRef] [PubMed]
8. Viceconti, M.; Taddei, F.; Van Sint Jan, S.; Leardini, A.; Cristofolini, L.; Stea, S.; Baruffaldi, F.; Baleani, M. Multiscale modelling of the skeleton for the prediction of the risk of fracture. *Clin. Biomech.* **2008**, *23*, 845–852. [CrossRef] [PubMed]
9. Murphy, S.V.; Atala, A. 3D bioprinting of tissues and organs. *Nat. Biotechnol.* **2014**, *32*, 773–785. [CrossRef] [PubMed]
10. Mota, C.; Puppi, D.; Chiellini, F.; Chiellini, E. Additive manufacturing techniques for the production of tissue engineering constructs. *J. Tissue Eng. Regen. Med.* **2015**, *9*, 174–190. [CrossRef] [PubMed]

applied sciences

MDPI

Article

Defining the URCOTEBS System as a Unilateral Radiographic–Stochastic Model for the Complementary States (Health/Disease) of the D-Organ and Middle-Ear Mucosa

Marian Rădulescu [1,†], Adela-Ioana Mocanu [2,*], Alexandra-Cristina Neagu [3,†], Mihai-Adrian Schipor [4] and Horia Mocanu [5,*,†]

1 Department of ENT&HNS, Faculty of Medicine, "Carol Davila" University of Medicine and Pharmacy, 020021 Bucharest, Romania
2 Department of ENT&HNS, Polimed Medical Center, 040067 Bucharest, Romania
3 Department of ENT&HNS, "Marie Sklodowska Curie" Emergency Children's Hospital, 041434 Bucharest, Romania; aaaneagu30@yahoo.com
4 Institute of Astronomical and Physical Geodesy, Technical University of Munich, 80333 Munich, Germany; mihai.schipor@unibw.de
5 Department of ENT&HNS, Faculty of Medicine, "Titu Maiorescu" University, 031593 Bucharest, Romania
* Correspondence: adela.mocanu@polimed.ro (A.-I.M.); horia.mocanu@prof.utm.ro (H.M.); Tel.: +40-723400435 (A.-I.M.); +40-732662268 (H.M.)
† These authors contributed equally to this work.

Abstract: The middle ear (ME) is a notoriously complicated anatomic structure, geometrically arranged as irregular interlinked spheroidal and polyhedric cavities dug inside the temporal bone (TB). The bony walls of these cavities are radiopaque and form the bony support for the *D-Organ* that we have previously defined as corresponding to the epithelium covering the Antrum walls (belonging to the central cavities of the middle ear) and the walls of mastoid and petrous cavities (the peripheral cavities of the ME). The aim of the study is to define an exact method for categorizing a **U**nilateral **R**adiographic **CO**nformation of the **TE**mporal **B**one in **S**chuller's projection (**URCOTEBS**) under one of the four defined conformations and using it for practical everyday clinical purposes. The conclusion is that a radiograph in Schuller's projection is a concrete way of storing precise information on the status (health/disease) of the *D-Organ* and therefore of the ME mucosa. These data is encoded within the image and we aim to decode and translate them into clinical data. The **URCOTEBS** results in an overlapping projection of all bony cavities that comprise the General Endo-temporal Bony Cavity Complex onto the same plain (film). This characteristic of classical film imaging constitutes an advantage from the multiple CT sections, as far as our proposed approach goes, because the set of stochastic information is found in the whole of the cavities taken as one on the same image, to which the measurement gauges can be easily applied. The decoding must be performed accordingly, and this occurs much faster with conventional radiography. This image of the TB in Schuller's projection is a mirror that reflects the status of the ME mucosa, and **URCOTEBS** encodes the physiological state of the *D-Organ*. The present work gives, through stochastic methods, the key to decoding this information into clinical language. In ascending order of their projection areas (projection of their *Variable Geometry Peripheral Endo-temporal Bony Cavity Complex*) we can recognize **URCOTEBS_d**, **URCOTEBS_c**, **URCOTEBS_b**, and **URCOTEBS_a**. The corresponding Greek letter designates the state of disease for each of these conformations: **URCOTEBS_δ**, **URCOTEBS_γ**, **URCOTEBS_β**, **URCOTEBS_α**, and the capital letters define their state of health: **URCOTEBS_D**, **URCOTEBS_C**, **URCOTEBS_B**, **URCOTEBS_A**. **URCOTEBS_d** is the smallest unilateral radiographic conformation of the TB in Schuller's projection and is, by definition, a radiographic image of the state of disease of the D-Organ. The probability of disease in **URCOTEBS_d** is 100%. This radiographic system is readily available and clinically usable.

Keywords: radiology; X-rays; tomography; otitis media; ventilation; anatomy; physiology; statistical modeling

1. Introduction

The *D-Organ* is a unistratified epithelial cellular population of homogenous morphology supported by a basilar membrane on top of a conjunctive structure that provides vascularization and innervation. The final layer of this conjunctive structure is the periosteum that binds it to the subjacent bone, which we will term the *Variable Geometry Anatomic Macroscopic Radiopaque Peripheral Support of the D-Organ*. This bony structure is geometrically arranged as irregular interlinked spheroidal and polyhedric cavities that form a common enclosure of completely determined volume V_ε [1,2]. The bony walls of these cavities form the aforementioned bony support. It is self-evident that these walls alone are dense enough to provide a radiographic image (for ease of description and writing we shall term this radiopaque); the cavities they define are radiolucent and contain a mix ε of σ types of respiratory gases saturated with water vapors representing a thermodynamic system ε^T completely defined by the three state parameters (p_ε, V_ε, T_ε) and delimited from its external environment (interface) by the unistratified epithelium called the *D-Organ* [1,2].

Each of these cavities communicates with at least six others (the six directions of space), which means that each cavity communicates, directly or indirectly, to all others thus forming the *Variable Geometry Peripheral Endo-temporal Bony Cavity Complex*, a term subservient to the larger term *Variable Geometry Peripheral Middle Ear* (Figure 1a,b). This complex communicates to the *Non-variable Geometry Central Axial Bony Cavities of the Middle Ear*, which in turn is subordinated to the notion of *Non-variable Geometry Central Axial Middle Ear*. The communication takes place through the corresponding orifices found in the walls of these central cavities (*Antrum, Additus, Protympanum, Tympanic cavity*)—Figure 2.

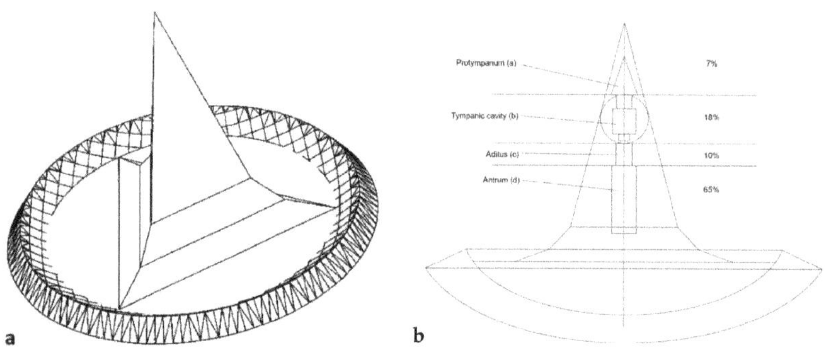

Figure 1. (**a**) Three-dimensional view of the regular composed geometric body representing the geometric–statistic model for the adult temporal bone [2]. (**b**) Longitudinal axial section through the geometric body representing the geometric–statistic model for the adult temporal bone. We can observe the position and fraction between the non-variable (central) and variable geometry middle ear (peripheral) [2].

The *F-Organ* is made up of a population of *F-Cells* whose *Anatomic Macroscopic Radiopaque Peripheral Support* is represented by the walls of the *Non-variable Geometry Central Axial Bony Cavities of the Middle Ear*, except the Antrum, which supports one of the *two D-Cell subpopulations* [1,2]. Thus, the cardinal of *F-Cell population* remains constant throughout a person's postnatal life and is two orders of magnitude smaller than the cardinal of *D-Cell population*. Practically, we will consider the middle-ear mucosa as a unit and the pathogenic and surface contribution of the *F-Cell population* insignificant by comparison to *D-Cell population* [1,2].

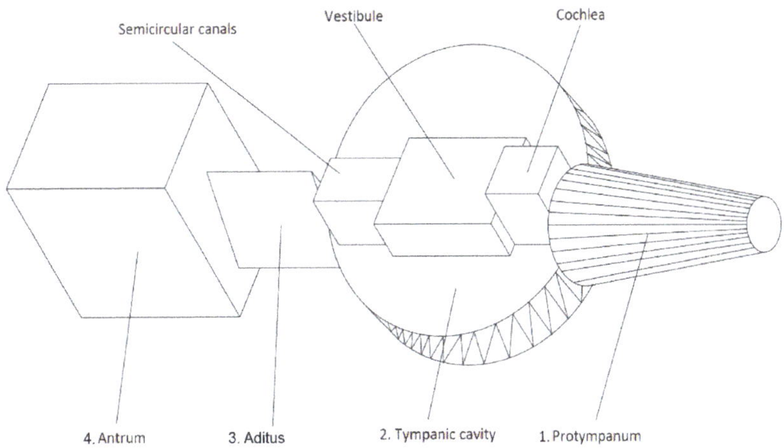

Figure 2. Three-dimensional representation of the fixed geometry middle ear and inner ear according to the geometric model of the temporal bone [2].

We can now define the *D-Organ* (general *D-Cell population*) as a sum of two subpopulations of *D-Cells*:

- *Central Axial D-Cell Subpopulation* of *constant* cardinal that has as *Anatomic Macroscopic Radiopaque Bone Support* the luminal surface of the Antrum—the size of this surface is constant throughout the subject's life starting from the seventh fetal month, and the cardinal of this subpopulation is constant and much smaller (by two orders of magnitude) than the other conspecific subpopulation.
- *Peripheral D-Cell Subpopulation* of *variable* cardinal that has as *Anatomic Macroscopic Radiopaque Bone Support* the variable geometry peripheral middle ear. The size of this surface grows or stops growing depending on the ratio of diseased *D-Cells*. The cardinal of this subpopulation is two orders of magnitude larger than the other conspecific subpopulation.

The *D-Cell* is a rectangular prism-shaped epithelial cell of minimal dimension (height h^D), perpendicular to the basilar membrane of the epithelium and the maximum dimension (l^D) parallel to the basilar membrane [1,2]—Figure 3.

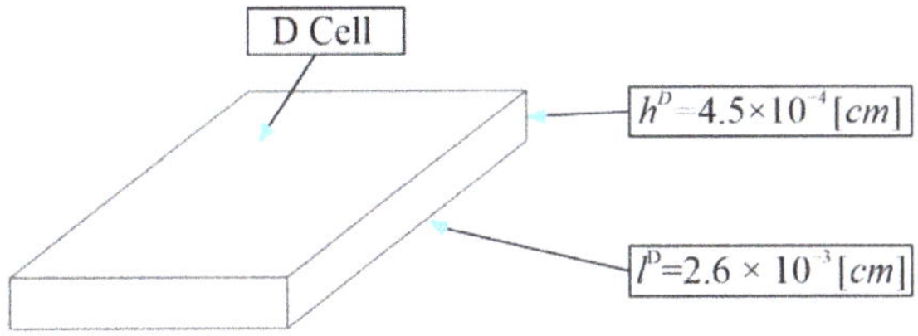

Figure 3. Statistic–geometric model for the D-Cell.

From a stochastic perspective, the *D-Cell* represents the statistical unit of the *D-Cell general population* (the cardinal of *D-Cells* comprising the *D-Organ*).

From a physiological point of view, each *D-Cell* is a functional unit of the *D-Organ* and it works as an energy consuming ionic membrane pump, working against trans-membrane and trans-cellular pressure gradients to transfer the three respiratory gas species (molecular oxygen, carbon dioxide, molecular nitrogen) from the internal environment of the human body into the middle-ear cavities. More precisely, there are three species of trans-membrane pumps corresponding to the three species of respiratory gases, all exclusively localized within the luminal membrane of the cell (one of the six surfaces of the cellular membrane) that faces the peripheral cavities of the middle ear (including the Antrum). This membrane is permeable to passive diffusion of respiratory gases (the *F*-Function* of the *D-Cell*) [1,2] according to Fick's law, similar to all membranes of the other human cells. The *F-Organ* fulfills the *F-Function*, which means the passive diffusion of respiratory gas molecules from the cavities of the ME into the internal environment of the body according to pressure gradients governed by Fick's law [1,2].

From a geometric, stochastic, and radiologic perspective, the *D-Organ* represents the sum of all *D-Cells* or the static general population of D-Cells. Axiomatically, we consider that the *D-Cell* can only be in one of the two complementary states: functional (Healthy) or non-functional (Diseased). We have previously stated that a halt in mastoid growth (*Variable Geometry Peripheral Endo-temporal Bony Cavity Complex*—Figure 2) occurs when the number of diseased *D-Cells* (*D-function abolished*) lays within the closed interval [0.28; 0.5]. The mastoid keeps growing as long as the number of diseased *D-Cells* is within the closed interval [0; 0.27]. The unilateral radiographic examination of the temporal bone (mastoid) in Schuller's projection shows the degree of development (growth) of the *Variable Geometry Peripheral Endo-temporal Bony Cavity Complex*, which is the *Variable Geometry Anatomic Macroscopic Radiopaque Peripheral Support of the D-Organ* and therefore the *Variable Geometry Anatomic Macroscopic Radiopaque Peripheral Support of the Middle-ear Mucosa.*

We would like to point out that this is osseous support and is defined by variable geometric sizes. These sizes are in themselves precise information on the status (health/disease) of the *D-Organ* and therefore of the ME mucosa. Thus, this bony structure is the material support for information encoded in its geometrical sizes and the radiographic image of the structure is the transcription of this stochastic information on film. The conclusion is that a radiograph in Schuller's projection is a concrete way of storing information on the status (health/disease) of the *D-Organ* and therefore of the ME mucosa. These data is encoded within the image and we aim to decode and translate them into clinical data.

The unilateral radiographic examination of the temporal bone (mastoid) in Schuller's projection results in an overlapping projection of all bony cavities that comprise the *General Endo-temporal Bony Cavity Complex* onto the same plain (film), and this constitutes an advantage from the multiple CT sections because the set of stochastic information is found in the whole of the cavities taken as one. The decoding must be performed accordingly, and this occurs much faster with conventional radiography, although other imagistic methods such as CT-scan and sometimes MRI are considered the standard in today's medicine.

2. Materials and Methods

In our previous article, we postulated that the state of the *D-Organ* (health/disease) can be statistically quantitatively defined, which gives the possibility of explaining the disease of the *D-Organ* (*Chronic Inflammation of the ME Mucosa*) through probability theory and mathematical statistics (stochastic).

We define chronic inflammation as a pathological status determined by irreversible intracellular lesions that annihilate the *D-Function* [1,2]. This is equivalent to total and permanent functional incapacity of the *D-Cell* exclusively in regard to the *D-Function*. All other cellular functions, including the *F*-Function*, remain normal. This means that

$$\text{Diseased D-Cell} \equiv \text{Non-functional D-Cell (permanent loss of D-Function)} \equiv \text{D-Cell with irreversible lesions} \equiv \text{D-Cell with chronic lesions.} \quad (1)$$

$$\text{Healthy D-Cell} \equiv \text{Functional D-Cell.}$$

As a consequence, the *Chronic Inflammatory Disease of the ME Mucosa* is equivalent to the disease status of the *D-Organ*, a status previously defined by quantitative statistical methods (1). We can state *D-Organ* disease when the specific weight of non-functional *D-Cells* is equal or higher than 28% but lower or equal to 50%. The principle behind this has been enunciated over 200 years ago by the French physician and surgeon *François Joseph*

Victor Broussais (1772–1838), who advocated the identity between physiology and pathology and that diseases are merely the effects of simple changes in intensity in the action of the stimulants that are indispensable for maintaining health [3,4]. This idea will be later on adopted by the philosopher and sociologist *Isidore Marie Auguste François Xavier Comte (1798–1857)*, who postulated that the identity of the normal and the pathological is affirmed as the acquisition of knowledge about the normal [4,5], and by the father of experimental physiology and founder of experimental medicine, *Claude Bernard (1813–1878)*, who, on the contrary, moves from the normal to the pathological with the aim of rational action directed at the pathological [4–6]. Broussais stated that all disease is an inflammation and irritation of tissues, especially of the digestive tract. He also invents the principle of identity between normal and pathologic, the only difference being of a quantitative nature. Acute inflammation is a totally reversible status that also may interest the *D-Organ* at times but only transitory and without permanent effects. There is no filiation between the two types of inflammations (reversible/irreversible–acute/chronic) and therefore there is no pathogenic linkage between the acute disease of the ME (acute otitis media) and chronic disease of the ME (chronic otitis media), which we especially term *Chronic Inflammatory Disease of the ME Mucosa* to make this distinction.

The fact that the status of the *D-Organ* is completely defined by two closed numerical intervals {[0; 0.27]; [0.28; 0.5]} representing the weight of *D-Cells* with irreversible lesions demonstrates that we are dealing with random mass phenomena that can be studied through probability theory and statistics. The essential characteristic of these phenomena is stability (consistency) of occurrence frequency.

The nature of the pathological agents and their interaction with the *D-Cells* interests us only to the extent where we can define the sum "*S*" of all conditions that ensure that the discussed events are *random*, *independent*, and *mass* events. The outside agent capable of producing a chronic lesion has, in itself, nothing special apart from the chance of interacting at a given time with the *D-Cells* (statistic units of a given population). By outside agent, we mean outside of the *D-Cell*, both surrounding environment and inside the subject's body. We will disregard the biological nature of the studied events (assailant agents, interaction between agents and cells, results of such interactions) and we will only calculate the odds (probability) of such events occurring (odds of assailant agents coming in contact with cells and odds of producing chronic irreversible lesions).

Question 1—What is the probability that an assailant agent comes into contact with the cells of the D-Organ? This is a complete set of events formed by two complementary (incompatible, opposing) events: E_1 = YES [comes into contact]; E_2 = NO [does not come into contact], which leads to the equality of the two probabilities $P(E_1) = P(E_2) = 1/2$. The two events form the certain event partitioning, which means the concomitant existence of the assailant agent and *D-Organ*.

Question 2—What is the probability that once in contact to the D-Cell the assailant agent interacts or does not interact with it? This is also a complete set of events similar to the previous, formed by two complementary (incompatible, opposing) events: E_3 = YES [it interacts]; E_4 = NO [does not interact], which leads to the equality of the two probabilities $P(E_3) = P(E_4) = 1/2$. This is the certain event partitioning, which means the potential existence of interaction between the assailant agent and *D-Cells* (implicitly *D-Function*).

Question 3—starting from the two previous answers, we propose the following problem:

Using the *unreplaced ball principle*, also known as *hypergeometric distribution* [7,8] (Figure 4), let us calculate the probability of obtaining random events with the following experiment: in an urn there are 50 black balls and 100 white balls (the ratio signifies the biological defense mechanisms of the D-Cell against any type of aggression—if the assailant agent interacts with the cell, the chances of irreversible damage are 50%); we extract 100 balls from the urn without returning any of them (*the unreturned ball*). Determine the probability of achieving a series of 25 independent, random events, meaning the pairs of integer numbers starting with 2 black balls–98 white balls and continuing with 4 black balls–96 white balls, 6 black balls–94 white balls . . . all the way to 50 black balls–50 white balls.

The unreplaced ball principle: if inside an urn there are balls of m colors: a_1 of color c_1; a_2 of color c_2; a_m of color c_m, and if we extract n balls concomitantly (or one by one without returning the extracted ball inside the urn), then the probability of obtaining α_1 balls of color c_1, α_2 balls of color c_2, α_m balls of color c_m ($\alpha_1+\alpha_2+\ldots+\alpha_m=n$) is:

$$\frac{C_{a_1}^{\alpha_1} C_{a_2}^{\alpha_2} \cdots \cdot C_{a_m}^{\alpha_m}}{C_{a_1+a_2+\ldots+a_m}^{\alpha_1+\alpha_2+\ldots+\alpha_m}}$$

Figure 4. The unreturned ball principle.

The number of black balls represents the weight of chronically diseased *D-Cells* (irreversible). This status is determined by the interaction between the assailant agent and the *D-cell* population and is assessed as such at the end of all contact between the two entities. The number of white balls represents the weight of healthy *D-Cells* (complementary), also assessed at the end of all contact between the two entities.

We have axiomatically conferred to the body's defense mechanism and therefore to the *D-Cell* also the intrinsic capacity to cut the chances of assailant success (irreversible lesions) in half: 100 white balls versus 50 black balls. This means that *Question 3* can be stated as follows: Determine the probability of achieving a series of 25 independent, random events completely defined by 25 pairs of integer numbers representing the weight of complementary diseased and healthy *D-Cells* ($\text{Weight}_{\text{diseased cells}} = 1 - \text{Weight}_{\text{healthy cells}}$). These 25 events represent the partition of the event E_1 = YES [comes into contact] with an already known probability of 1/2 and 25 possible manners of intensity of said aggression. These probabilities are valid for a single *D-Cell* population forming a single *D-Organ* located in the temporal bone of a single individual. To obtain information regarding the general human population or a segment thereof, the corresponding temporal bone population (one-side), and the corresponding *D-Organ* population for these TBs, we must calculate the probabilities of achieving each of these 25 events this time considered as Exclusive Random Events. We shall use Poisson's scheme [9,10] (Figure 5). In other words, we must calculate the probability of having one of the 25 Independent Random Events without having any of the other (see Table 1).

Poisson's scheme

If independent events $A_1, A_2, \ldots A_n$ have known probabilities: $P(A_1) = p_1$; $P(A_2) = p_2$; $P(A_n) = p_n$; then the probability of having k events out of the n (and not have $n-k$) is the coeficient x^k from the polinomyal $Q(x)=(p_1x + q_1)(p_2x+q_2)\ldots(p_nx+q_n)$; where $q_i=1-p_i$; $i=1, 2, \ldots n$. The probabily of this event occuring is:

$$\Sigma p_{i_1} \cdot p_{i_2} \cdots \cdot p_{i_k} q_{i_{k+1}} \cdots \cdot q_{i_n}$$

Figure 5. Poisson's scheme.

Table 1. Probabilities of achieving the 25 Random Exclusive Events calculated with the unreturned ball scheme and Poisson's scheme.

Event Index	White Balls (Weight of Healthy D-Cells)	Black Balls (Weight of Diseased D-Cells)	Event	Primitive Probability (Unreturned Ball Scheme and Poisson's Scheme)	Corrected Primitive Probability (Brought to 1/2)
1	2	3	4	5	6
1	98	2	98_2	2.46×10^{-10}	1.23×10^{-10}
2	96	4	96_4	5.67×10^{-9}	2.84×10^{-9}
3	94	6	94_6	8.22×10^{-8}	4.11×10^{-8}
4	92	8	92_8	8.41×10^{-7}	4.20×10^{-7}
5	90	10	90_10	6.47×10^{-6}	3.23×10^{-6}
6	88	12	88_12	3.90×10^{-5}	1.95×10^{-5}
7	86	14	86_14	1.89×10^{-4}	9.46×10^{-5}
8	84	16	84_16	7.54×10^{-4}	3.77×10^{-4}
9	82	18	82_18	2.50×10^{-3}	1.25×10^{-3}
10	80	20	80_20	7.02×10^{-3}	3.51×10^{-3}
11	78	22	78_22	1.67×10^{-2}	8.36×10^{-3}
12	76	24	76_24	3.42×10^{-2}	1.71×10^{-2}
13	74	26	74_26	6.03×10^{-2}	3.01×10^{-2}
14	72	28	72_28	9.21×10^{-2}	4.61×10^{-2}
15	70	30	70_30	1.22×10^{-1}	6.12×10^{-2}
16	68	32	68_32	1.42×10^{-1}	7.10×10^{-2}
17	66	34	66_34	1.44×10^{-1}	7.21×10^{-2}
18	64	36	64_36	1.28×10^{-1}	6.41×10^{-2}
19	62	38	62_38	1.00×10^{-1}	5.01×10^{-2}
20	60	40	60_40	6.88×10^{-2}	3.44×10^{-2}
21	58	42	58_42	4.15×10^{-2}	2.08×10^{-2}
22	56	44	56_44	2.20×10^{-2}	1.10×10^{-2}
23	54	46	54_46	1.03×10^{-2}	5.15×10^{-3}
24	52	48	52_48	4.22×10^{-3}	2.11×10^{-3}
25	50	50	50_50	1.52×10^{-3}	7.62×10^{-4}
	Sum of calculated probabilities			**100%**	**50%**

The complementary event to E_1 is E_2 = NO [does not come into contact], also with a 1/2 probability. This one is equivalent to the 100% weight of healthy *D-Cells* and also 0% weight of diseased *D-Cells*. To facilitate this, the last event we discuss had the index 0 and was called Event 100_0 (100% white balls, 0% black balls)—see Table 2. This finally defines a complete system of 26 Exclusive Random Events of known probabilities (Table 2).

Table 2. Probabilities of the complete 26 Exclusive Random Events system.

Event Index	White Balls (Weight of Healthy D-Cells)	Black Balls (Weight of Diseased D-Cells)	Event	Primitive Probability (Unreturned Ball Scheme and Poisson's Scheme)	Corrected Primitive Probability (Brought to 100%)
1	2	3	4	5	6
0	100	0	100_0	5.00×10^{-1}	5.00×10^{-1}
1	98	2	98_2	2.46×10^{-10}	1.23×10^{-10}
2	96	4	96_4	5.68×10^{-9}	2.84×10^{-9}
3	94	6	94_6	8.24×10^{-8}	4.12×10^{-8}
4	92	8	92_8	8.43×10^{-7}	4.21×10^{-7}
5	90	10	90_10	6.48×10^{-6}	3.24×10^{-6}

Table 2. *Cont.*

Event Index	White Balls (Weight of Healthy D-Cells)	Black Balls (Weight of Diseased D-Cells)	Event	Primitive Probability (Unreturned Ball Scheme and Poisson's Scheme)	Corrected Primitive Probability (Brought to 100%)
6	88	12	88_12	3.91×10^{-5}	1.95×10^{-5}
7	86	14	86_14	1.89×10^{-4}	9.47×10^{-5}
8	84	16	84_16	7.55×10^{-4}	3.78×10^{-4}
9	82	18	82_18	2.50×10^{-3}	1.25×10^{-3}
10	80	20	80_20	7.03×10^{-3}	3.52×10^{-3}
11	78	22	78_22	1.67×10^{-2}	8.37×10^{-3}
12	76	24	76_24	3.43×10^{-2}	1.71×10^{-2}
13	74	26	74_26	6.04×10^{-2}	3.02×10^{-2}
14	**72**	**28**	**72_28**	$\mathbf{9.23 \times 10^{-2}}$	$\mathbf{4.61 \times 10^{-2}}$
15	70	30	70_30	1.22×10^{-1}	6.11×10^{-2}
16	68	32	68_32	1.42×10^{-1}	7.11×10^{-2}
17	66	34	66_34	1.44×10^{-1}	7.21×10^{-2}
18	64	36	64_36	1.28×10^{-1}	6.41×10^{-2}
19	62	38	62_38	1.00×10^{-1}	5.01×10^{-2}
20	60	40	60_40	6.89×10^{-2}	3.45×10^{-2}
21	58	42	58_42	4.16×10^{-2}	2.08×10^{-2}
22	56	44	56_44	2.20×10^{-2}	1.10×10^{-2}
23	54	46	54_46	1.03×10^{-2}	5.16×10^{-3}
24	52	48	52_48	4.23×10^{-3}	2.11×10^{-3}
25	50	50	50_50	1.52×10^{-3}	7.61×10^{-4}
	Sum of calculated probabilities			1.50	1.00

We notice that the first 14 events in Table 2 (0, 1, 2, . . ., 13) correspond to the definition of Healthy *D-Organ* and the last 12 to the Diseased *D-Organ* [1,2]. Event number 14 represents the minimum weight of diseased cells, starting from which the growth of the *D-Organ* comes to a halt [2]. This event is the key to explaining the entire phenomena. The status of the *D-Organ* corresponding to event 14 (Exclusive Event 72_28) will be completely determined if it can be identified with its physical (geometric) and functional (metabolic energy consumption of a *D-Cell*, q_i) status at the time of birth (the *D-Organ* becomes functional starting from the seventh fetal month) [1].

During intra-uterine life, any of the abovementioned events is axiomatically considered as *impossible events* (probability 0). If we admit this postulate, then the moment of birth is defined as a benchmark (numerical value 1) for the functional status (metabolic energy consumption necessary for *D-Function*) and anatomic (number of *D-Cells*) status, reflected by the geometric measures of the *Endo-temporal Bony Cavity Complex* and the *Anatomic Macroscopic Radiopaque Bony Support of the D-Organ*. In other words, we confer the numerical value 1 to the two characteristics of the *D-Organ* considered at the moment of birth: metabolic energy consumption and geometric measurements.

Starting from this and based on *Geometric–statistical model for middle-ear anatomy and ventilation* [2] (Figure 6) and the *Thermodynamic Model for Middle-Ear Ventilation* [1], we can determine a matrix based on the data from the 26 Exclusive Random Events with their corresponding probabilities (Table 3). The connection between this data matrix and practice (clinical) is the unilateral radiograph of the TB under Schuller's projection (the geometrical measurements belonging to the said matrix)—see Table 3, column 8.

The starting point for this connection is the axiomatic definition of the equivalency between the series of the last 12 events (E14–E25) in the data matrix to any experimental series of the smallest unilateral radiographic images of the *Endo-temporal Bony Cavity Complex* and the *Anatomic Macroscopic Radiopaque Bony Support of the D-Organ* or of the smallest unilateral radiographic conformations of the TB in Schuller's projection (URCOTEBS), which we will note as **URCOTEBS_d**. This series of events completely defines the state of disease.

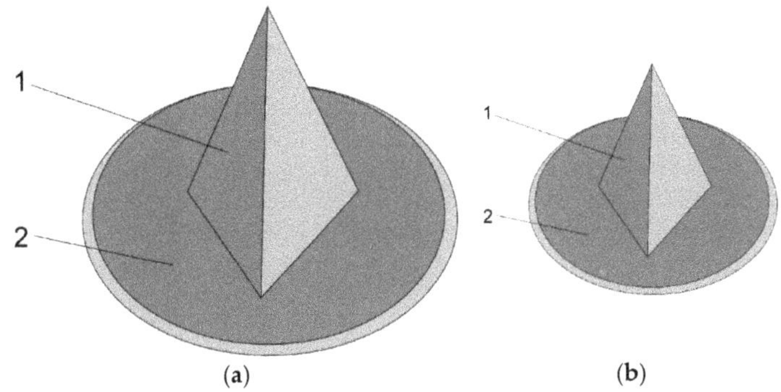

Figure 6. (**a**) Three-dimensional representation of the temporal bone according to the geometric model for an adult: 1—right pyramid with a triangular base; 2—spherical cap. (**b**) Three-dimensional representation of the temporal bone according to the geometric model for a newborn: 1—right pyramid with a triangular base; 2—spherical cap.

Table 3. Data matrix.

Event Index	Event	Unilateral Radiographic Conformation of the Temporal Bone in Schuller's Projection	White Balls (Weight of Healthy D-Cells) [%]	Black Balls (Weight of Diseased D-Cells) [%]	Corrected Primitive Probability (from the Unreturned Ball Scheme and Poisson's Scheme) (Brought to 100%)	Energy Consumption Relative to Consumption at the Time of Birth	Total Volume of Pneumatized Cavities Relative to the Volume at the Time of Birth
1	2	3	4	5	6′	7	8
0	100_0	a	100	0	5.00×10^{-1}	3%	6.00
1	98_2	a	98	2	1.23×10^{-10}	6%	5.64
2	96_4	a	96	4	2.84×10^{-9}	9%	5.29
3	94_6	a	94	6	4.12×10^{-8}	12%	4.93
4	92_8	a	92	8	4.21×10^{-7}	15%	4.57
5	90_10	a	90	10	3.24×10^{-6}	18%	4.21
6	88_12	b	88	12	1.95×10^{-5}	21%	3.86
7	86_14	b	86	14	9.47×10^{-5}	24%	3.50
8	84_16	b	84	16	3.78×10^{-4}	28%	3.14
9	82_18	b	82	18	1.25×10^{-3}	31%	2.79
10	80_20	b	80	20	3.52×10^{-3}	34%	2.43
11	78_22	c	78	22	8.37×10^{-3}	37%	2.07
12	76_24	c	76	24	1.71×10^{-2}	40%	1.71
13	74_26	c	74	26	3.02×10^{-2}	43%	1.36
14	72_28	d	72	28	4.61×10^{-2}	100.00%	1
15	70_30	d	70	30	6.11×10^{-2}	103%	1
16	68_32	d	68	32	7.11×10^{-2}	106%	1
17	66_34	d	66	34	7.21×10^{-2}	109%	1
18	64_36	d	64	36	6.41×10^{-2}	112%	1
19	62_38	d	62	38	5.01×10^{-2}	115%	1
20	60_40	d	60	40	3.45×10^{-2}	118%	1
21	58_42	d	58	42	2.08 × 10−2	121%	1
22	56_44	d	56	44	1.10 × 10−2	124%	1
23	54_46	d	54	46	5.16 × 10−3	128%	1
24	52_48	d	52	48	2.11 × 10−3	131%	1
25	50_50	d	50	50	7.61 × 10−4	134%	1
		Sum of calculated probabilities			1.00		

The matrix also completely defines the other three types of unilateral radiographic conformations of the TB in Schuller's projection (URCOTEBS), which we termed in ascending order of their projection areas (projection of their *Variable Geometry Peripheral Endo-temporal Bony Cavity Complex*): **URCOTEBS_c, URCOTEBS_b, URCOTEBS_a**. This definition is, after all, the concatenation of the matrix data based on the scaling theory and classification theory. Also note that we used the corresponding Greek letter to designate the state of disease for each of these conformations: **URCOTEBS_δ, URCOTEBS_γ, URCOTEBS_β, URCOTEBS_α**, and capital letters for their state of health: **URCOTEBS_D, URCOTEBS_C, URCOTEBS_B, URCOTEBS_A**.

Table 4 presents the complete definitions of the four radiographic conformations, each representing an Exclusive Random Event and having its own probability. The stability (consistency) of this probability is the essential characteristic of Random Mass Events. By mass, we understand both the *D-Cell population* and also any general human population. Thus, Random Mass Events have a defining general tendency toward constancy of probability achievement.

Table 4. Data matrix concentrated on defining the unilateral radiographic conformations of the TB in Schuller's projection through mean numerical values of the energy and volume contribution, as well as the weight of healthy and diseased D-Cells.

Unilateral Radiographic Conformations of the TB in Schuller's Projection (URCOTEBS)	Event Index	White Balls (Weight of Healthy D-Cells) [%]	Black Balls (Weight of Diseased D-Cells) [%]	Corrected Primitive Probability (from the Unreturned Ball Scheme and Poisson's Scheme) (Brought to 100%)	Mean Energy Consumption Relative to Consumption at the Time of Birth (q_i)	Total Volume of Pneumatized Cavities Relative to the Volume at the time of Birth (v_i)
1′	2′	3′	4′	5′	6	7′
a	0–5	95	5	5.97×10^{-1}	10.72%	5.11
b	6–10	84	16	7.54×10^{-3}	27.56%	3.14
c	11–13	76	24	1.33×10^{-1}	39.80%	1.71
d	14–25	61	39	2.62×10^{-1}	108.51%	1
	Sum of calculated probabilities			1.00		

According to the law of large numbers (Moivre–Laplace), the frequency of general prevalence of the four unilateral radiographic conformations of the TB tends to be always the same for any given general human population (Table 5). Also see Table 6 for visual reference.

Table 5. The probabilities of Exclusive Random Events achievement constituted by the four Unilateral Radiographic Conformations (URCOTEBS): **URCOTEBS_d; URCOTEBS_c; URCOTEBS_b; URCOTEBS_a**.

Event Index	White Balls (Weight of Healthy D-Cells) [%]	Black Balls (Weight of Diseased D-Cells) [%]	Corrected Primitive Probability (from the Unreturned Ball Scheme and Poisson's Scheme) (Brought to 100%)
2′	4′	5′	6′
0–5	95	5	5.97×10^{-1}
6–10	84	16	7.54×10^{-3}
11–13	76	24	1.33×10^{-1}
14–25	61	39	2.62×10^{-1}
	Sum of calculated probabilities		1.00

Table 6. Schematic images of each **URCOTEBS** conformation.

	Health	Disease
a	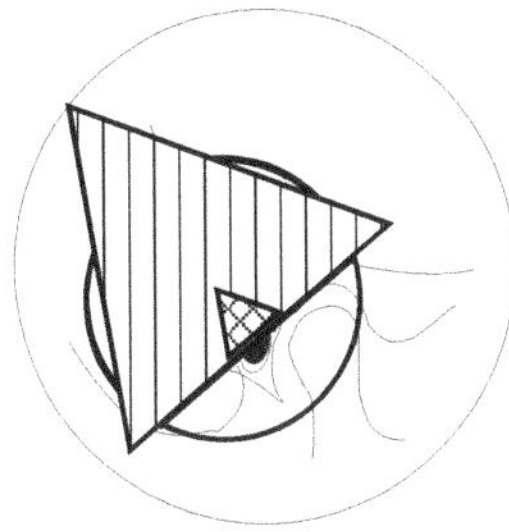 **A** P€ = 99.4%	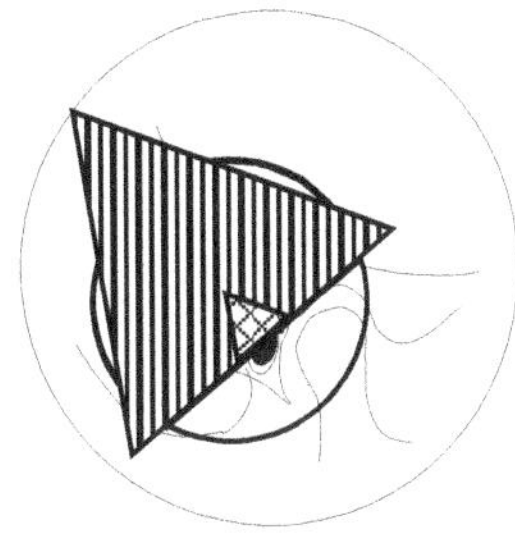 α P€ = 0.6%
b	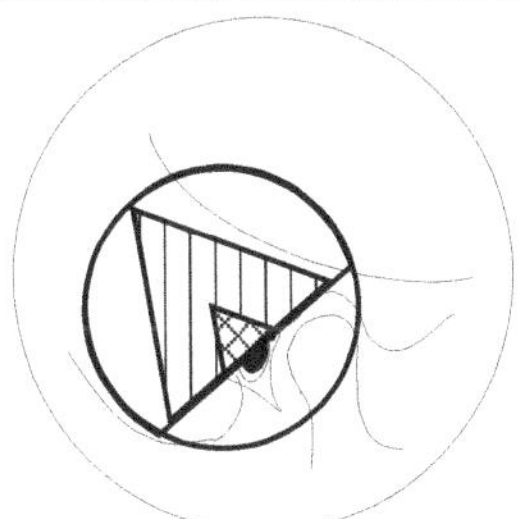 **B** P€ = 98.6%	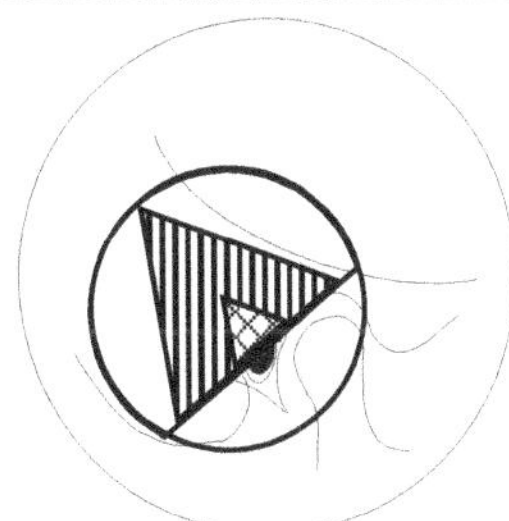 β P€ = 1.4%
c	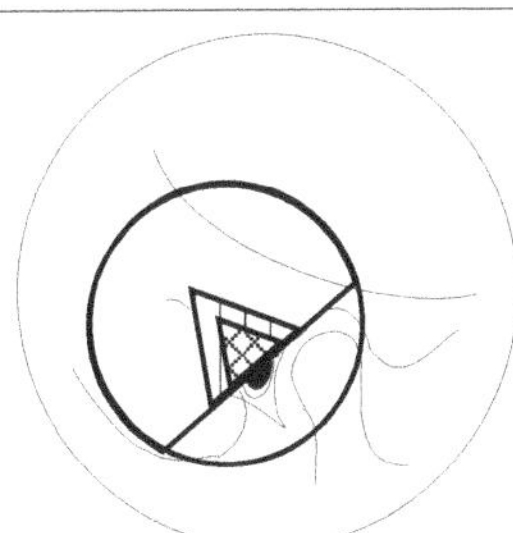 **C** P€ = 92.1%	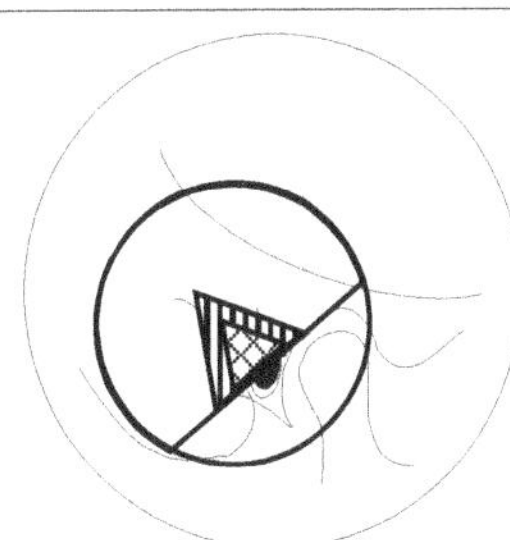 γ P€ = 7.9%
d	Imposible event P€ = 0%	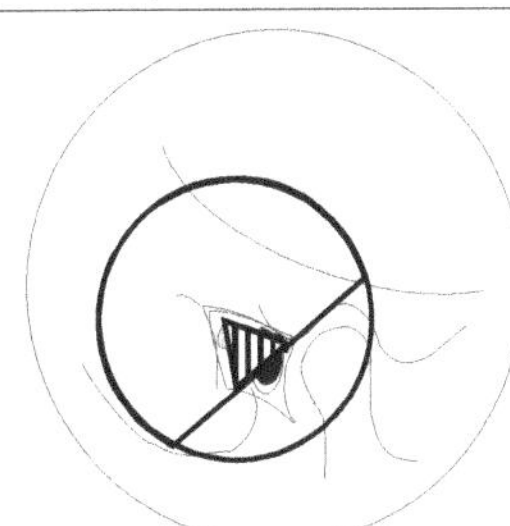 δ Certain event P€ = 100%

3. Results

The aim of the study is to define an exact method for categorizing a unilateral radiographic scan of the TB (mastoid) in Schuller's projection (URCOTEBS) under one of the four conformations and using it for practical everyday clinical purposes.

The otic capsule (OC) achieves its maximum volume during the seventh month of fetal life and from the moment of birth has a maximum 2 mm. appositional growth only. In other words, if we consider the OC as a spherical object, the adult sphere is only 2 mm. larger than the newborn sphere. Based on these considerations, we can define the radiographic opacity of the OC as the gauge for measuring the projection area of the *Variable Geometry Peripheral Endo-temporal Bony Cavity Complex* or the *Anatomic Macroscopic Radiopaque Bony Support of the D-Organ.* For achieving graphic accuracy, we will circumscribe the radiographic opacity of the OC inside an equilateral triangle as a measuring reference. This is the most internal triangle of the orthogonal axial projection of the base of the regular composite geometric figure that is the geometric model of the TB in newborns and adults. This equilateral triangle has practically the same area in both newborns and adults. In Figure 7, we notice how the projection of the Non-variable Geometry Central Axial Bony Cavities of the Middle ear is inscribed within the triangle (Also see Figures 1 and 2).

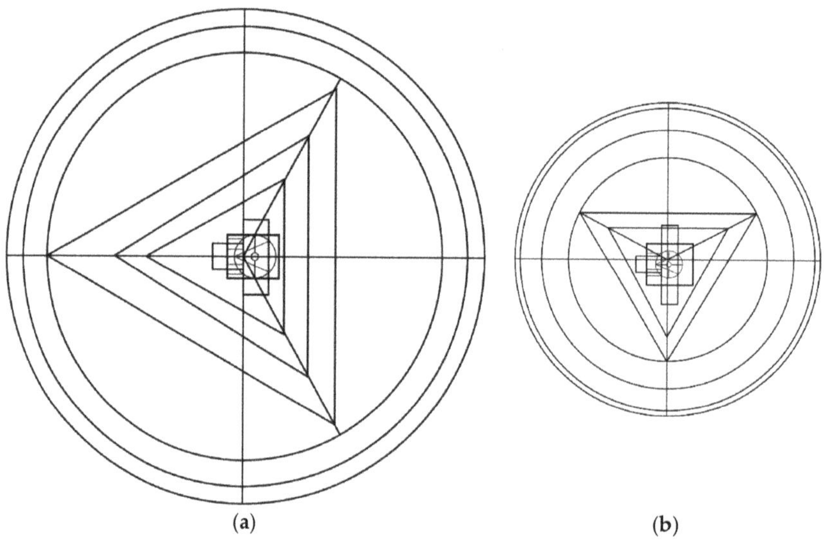

Figure 7. Axial orthogonal projections of the base of regular composed geometrical bodies representing the geometric–statistic model of the adult (**a**) and newborn (**b**) temporal bone. We can observe that the Central Axial Non-variable Geometry of the middle ears of adults and newborns has practically the same shape and size and that the equilateral triangles circumscribing it and the otic capsule are of the same areas.

The following graphic representations will depict the external circle of the orthogonal projection of the basis of the geometric figure in adults and the inner most equilateral triangle of the projection in newborns (Figure 7). We will also use a corresponding semicircle and the drawn schematics of a Schuller radiograph inscribed within the initial circle. By overlapping the three figures, namely the circle, semicircle, and drawing (Figure 8), we will establish the natural position of the initially described triangle. This triangle inscribes the projection of the following structures on radiographic film: OC opacity (bony labyrinth), non-variable geometry central axial middle ear (non-variable geometry cavities), coaxial transparency of the internal and external acoustic meatus. (Figures 7–12).

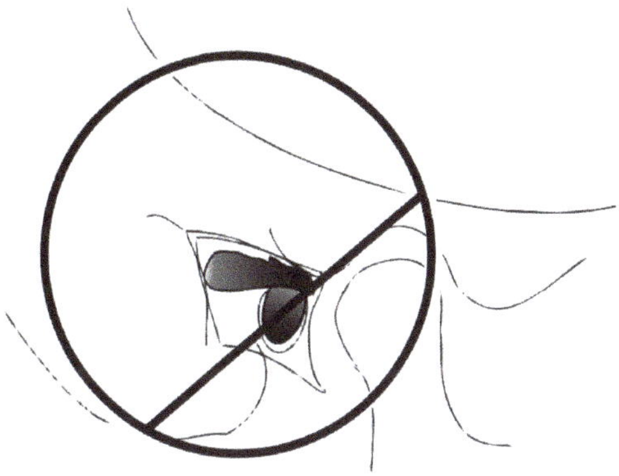

Figure 8. The schematic representation of an URCOTEBS radiograph with the circle and semicircle of the base of the orthogonal projection of the geometric body resulting from modelling the adult ME.

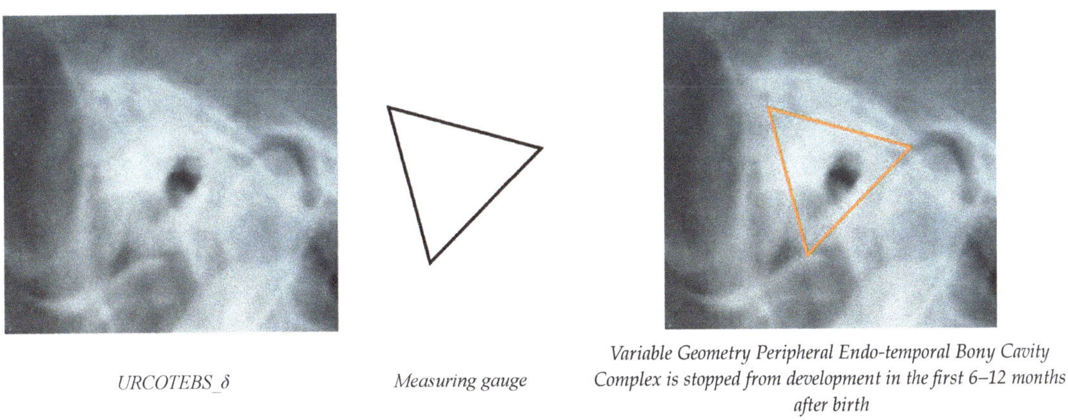

Figure 9. URCOTEBS_δ is pathognomonic for the state of disease of the D-Organ, equivalent to the state of disease of the ME mucosa − Certain event P(E) = 100%.

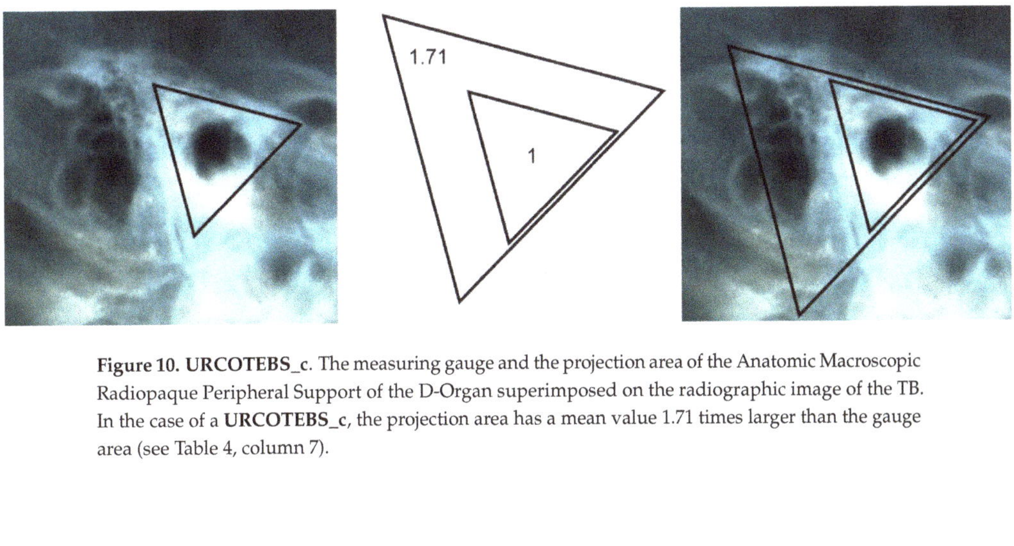

Figure 10. URCOTEBS_c. The measuring gauge and the projection area of the Anatomic Macroscopic Radiopaque Peripheral Support of the D-Organ superimposed on the radiographic image of the TB. In the case of a **URCOTEBS_c**, the projection area has a mean value 1.71 times larger than the gauge area (see Table 4, column 7).

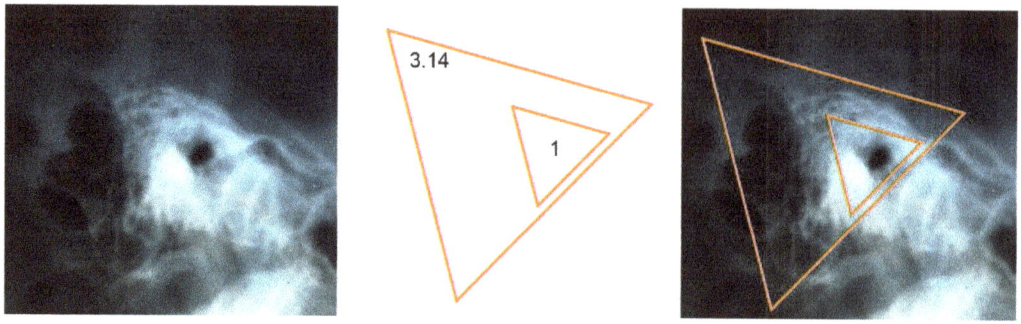

Figure 11. URCOTEBS_b. The measuring gauge and the projection area of the Anatomic Macroscopic Radiopaque Peripheral Support of the D-Organ superimposed on the radiographic image of the TB. In the case of a **URCOTEBS_b**, the projection area has a mean value 3.14 times larger than the gauge area (see Table 4, column 7).

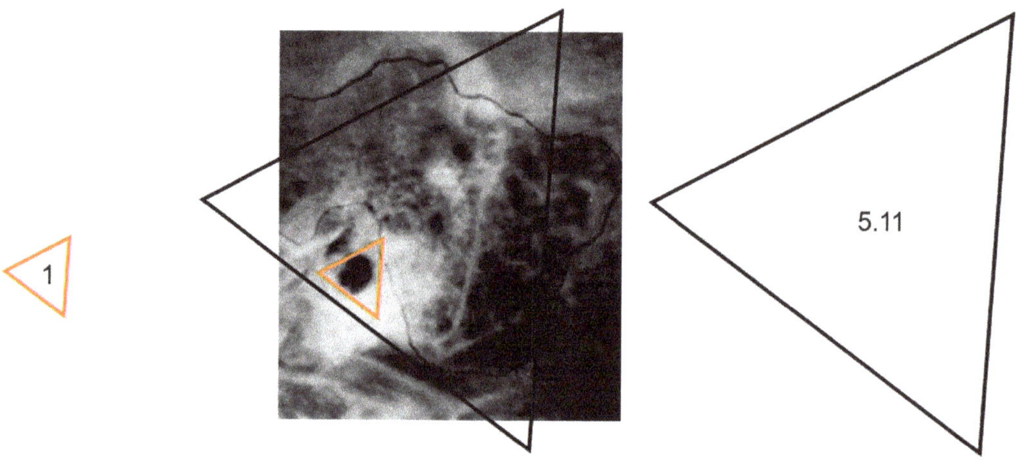

Figure 12. URCOTEBS_a. The measuring gauge and the projection area of the Anatomic Macroscopic Radiopaque Peripheral Support of the D-Organ superimposed on the radiographic image of the TB. In the case of a **URCOTEBS_a**, the projection area has a mean value 5.11 times larger than the gauge area (see Table 4, column 7).

In this manner, we can describe schematic images of radiographs, analog icons, stylized icons, all of which represent the *unilateral radiographic conformations of the TB in Schuller's projection (URCOTEBS)*. See Tables 6–8.

From the previous paragraphs, we conclude that the equilateral triangle as a measuring gauge is implicitly present in any radiograph of any patient, and we only need to draw it upon the radiographic film (Figures 9–12). Thus, any radiograph of any patient will have its own measuring triangle and we will use a reference scale to measure the area of the *Anatomic Macroscopic Radiopaque Bony Support of the D-Organ*. This represents an enormous practical advantage. In other words, each image contains in itself the first division (0–1) and the first point (the equilateral triangle of point 1) of a unique measuring scale (reference scale). This means that the radiographic projection of the *Variable Geometry Peripheral Middle Ear* to be measured with this scale is outside the triangle gauge perimeter. We will draw next the contour of the *Variable Geometry Peripheral Middle Ear*, and this will be inscribed within an equilateral triangle larger than the gauge triangle as shown in Figures 9–12. Finally, we calculate the ratio between the area of the gauge and the area of the projection triangle. Based on this numerical value and the values in the data matrix (Table 3, column 8 and Table 4, column 7), we can fit each unilateral radiographic image within one of the previously defined conformations (URCOTEBS_d, URCOTEBS_c, URCOTEBS_b, URCOTEBS_a).

Table 7. Analog icons for **URCOTEBS**.

	Health Icons	**Disease Icons**
	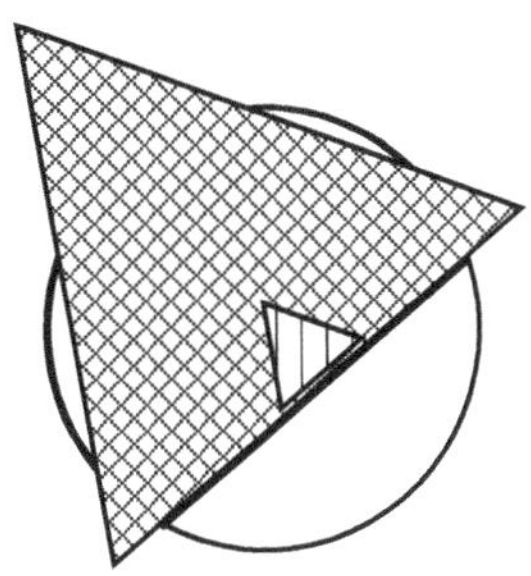	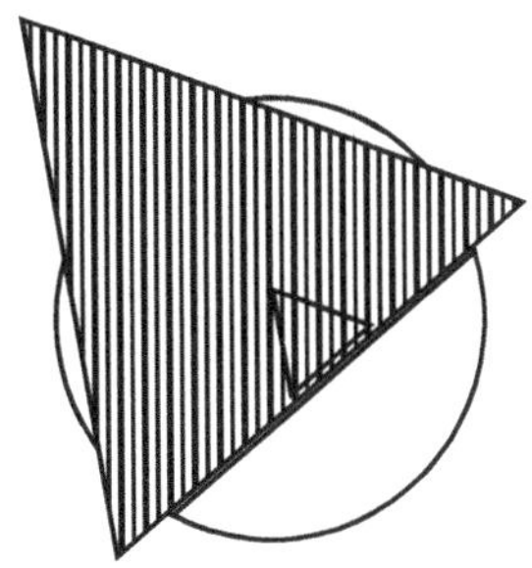
a	A	α
	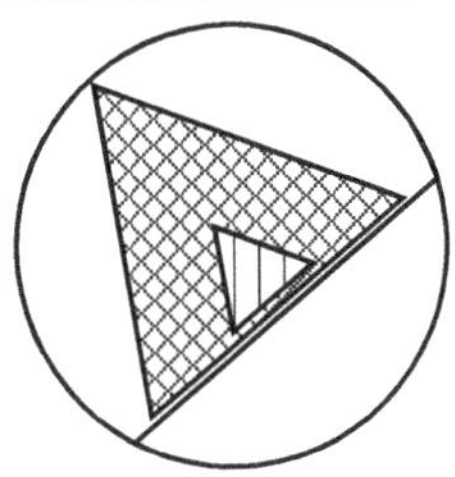	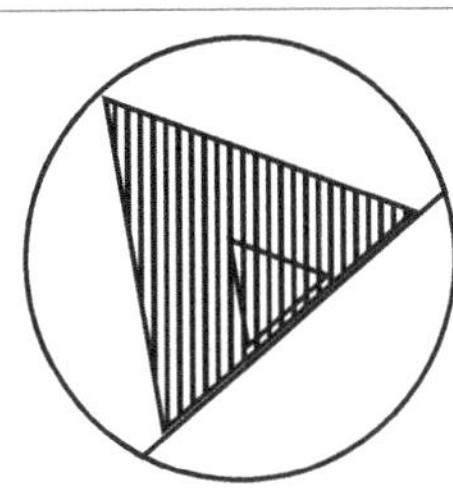
b	B	β
	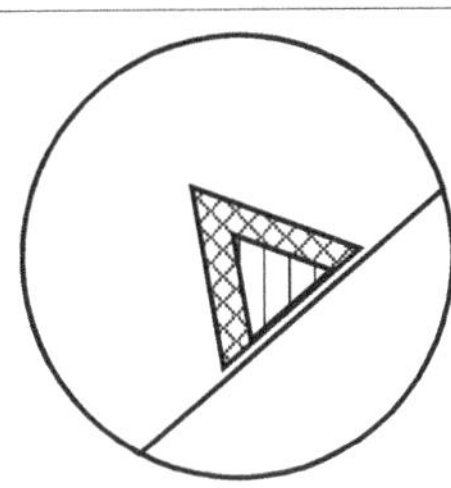	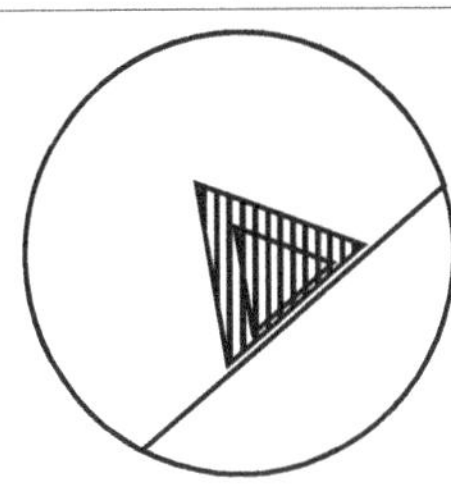
c	C	γ
	Imposible Random Exclusive Event	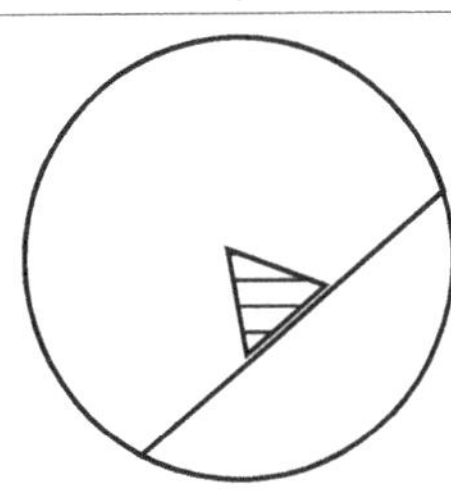
d	non	δ

Table 8. Stylized icons for **URCOTEBS**.

	Health	Disease
	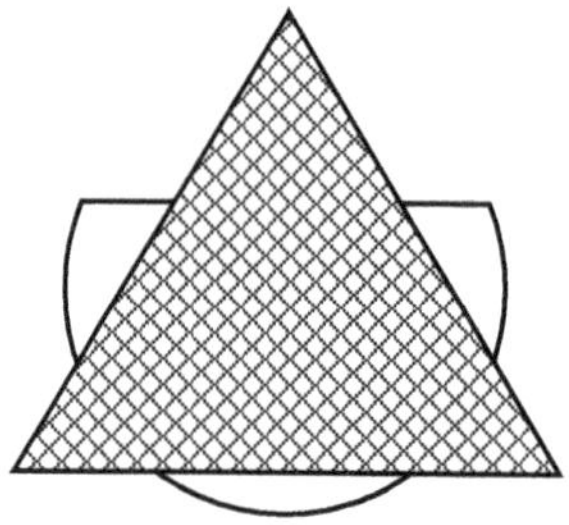	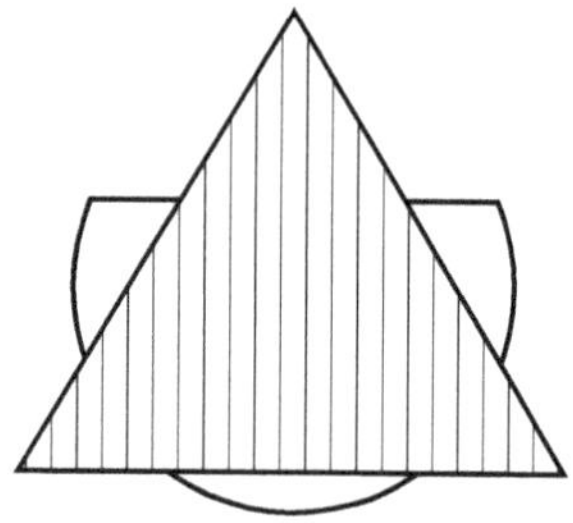
a	A	α
	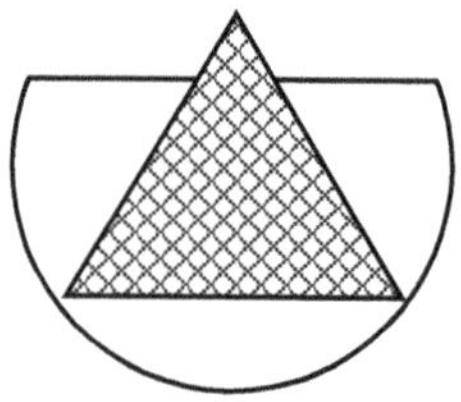	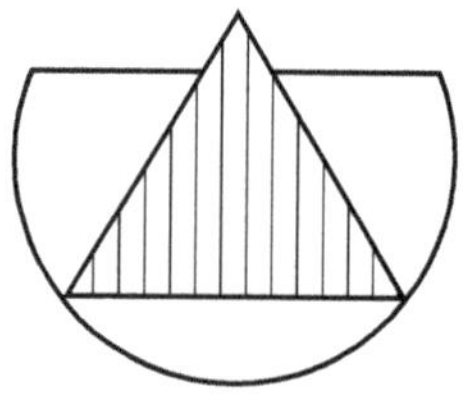
b	B	β
	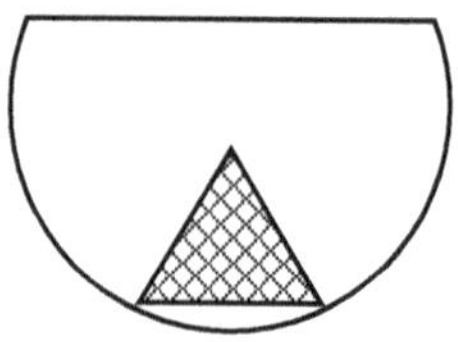	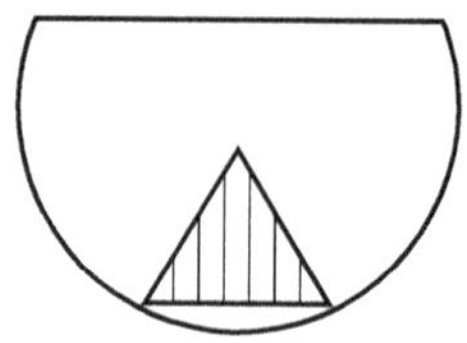
c	C	γ
	Imposible Random Exclusive Event	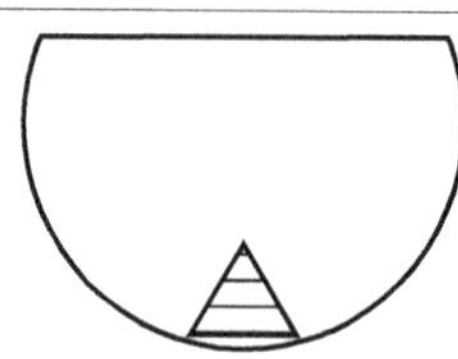
d	Imposible event	δ

In this manner, we can draw the following: schematic images of the radiographs of the four URCOTEBS (see Table 6), analog icons for URCOTEBS (Table 7), stylized icons for URCOTEBS (Table 8).

4. Discussions

Starting from the conclusions that we already presented in our previous articles, we can safely state that the growth of the mastoid stops when the weight of diseased *D-Cells* is equal or higher than 28% [1,2]. Based on this, we can determine the growth rate of the *D-Organ* (Table 3, column 8 and Table 4, column 7). This growth is an analytical function of the weight of diseased *D-Cells*. Determining this ratio was achieved by defining the level of anatomic (geometric) development of the *D-Organ* at the moment of birth by the numerical value 1 and simply applying the rule of three.

Initially, only four numerical values for the weight of diseased *D-Cells* were determined {[0; 0.27]; [0.28; 0.5]} representing the limits of two variation intervals. These intervals have a very precise significance by defining the states of health and disease of the *D-Organ* [2]. The other previously undetermined values of the weight of diseased cells have been completely defined by approaching the pathology problem exclusively from a probability theory point of view.

We must now decode the statistic and probabilistic information encrypted in each type of **URCOTEBS** conformation. Each radiograph is encrypting statistic and stochastic data regarding the status of the *D-Organ*. We consider **URCOTEBS** a mirror that never lies about the *Danaid Organ*.

URCOTEBS_d is, by definition, a radiographic image of the state of disease of the *D-Organ* [2]. Thus, the probability of disease in a **URCOTEBS_d** is 100% [1]. This is a certain event probability [2], and this conformation is pathognomonic for *Chronic Inflammation of the ME Mucosa*. We must now determine the homologous probabilities for the other three conformations. The answer lies in the conclusion on the reasons for mastoid growth [2]. It is a question of metabolic energy economy for performing the *D-Function*. Once saved, it will be redistributed to defense mechanisms and other cellular function that require metabolic energy. This means that the probability of irreversible lesion is an analytical function of the energy consumption required for performing the *D-Function*. Thus, mastoid growth means increasing chances of maintaining health status [2].

A mathematical determination of the energy consumption reduction rate is identical to determining the volume growth rate. The data matrix (Table 3, column 6) contains numerical values for these determinations. Table 4 contains the processed values for the same determinations. Based on these values, we calculate the probability for disease presence for each of the three conformations—**URCOTEBS**_c, **URCOTEBS**_b, **URCOTEBS**_a. The mathematical relations for these calculations (Equations (1)–(7)) are based on the rule of three. It is self-explanatory that, based on complementarity, we implicitly also determined the probability for health status (Tables 9 and 10, Equations (1)–(7)).

Middle-ear pathology leading to hearing loss is the most serious problem confronting the otologist, as the WHO estimates that between 65 and 330 million individuals suffer from some form of middle ear suppuration and that 50% of them suffer from hearing impairment, which could also be caused by a number of clinical and environmental factors [11–13]. The middle-ear or tympanic cavity is a very complicated structure, with notoriously intricate small-sized segments geometrically arranged as irregular interlinked spheroidal and polyhedric cavities. The geometry of the mastoid air cells system (MACS) has not been described for adults over a large range of MACS volumes [14], but numerous studies reports that these volumes are indirectly related to predisposition for middle-ear pathology such as chronic otitis media [15–19]. Measuring the size and pneumatization of the mastoid bone has been a current issue with otologists, as MACS functions as a gas reserve for the ME and slows the rate of middle-ear pressure decrease between Eustachian tube openings [20]. This would mean that MACS and not the Eustachian tube is the primary gas source for the ME [1,2,21]. The methods for measuring the volumes include the water-weight method, the pressurized transducer method, and planimetric methods [22–24].

The volume of interconnected air cells is difficult to measure, which is why various methods have been employed: researchers such as Palva et al., Csakanyi et al., and Valtonen et al., use planimetric methods [19,25–28], while others such as Koç et al., Vrabec et al., and Lee et al., advocate for volumetric methods [24–26,29–32]. An article by Park et al., reported that the MACS surface area is a linear function of the MACS volume [23]. This is also confirmed by the studies of Swarts et al., and Colhoun et al. [14,33], who used Schuller's projection X-rays and reconstructions from CT sections to document this linear relationship with a high correlation coefficient ($r \approx 0.77$ and $r \approx 0.95$, respectively). The conclusion of all this research is that relatively simple X-ray procedures can be used to screen ears for the MACS volume [14].

$$f(\alpha) = 1 - f(A) = \frac{1}{1 + \sum_{t=1}^{5} \lambda \overline{q}_t(d;\Gamma_t) + \sum_{t=1}^{5} \overline{q}(a)\overline{q}_t(d;\Gamma_t)} \quad (1)$$

$$f(\beta) = 1 - f(B) = \frac{1}{1 + \sum_{t=1}^{5} \overline{q}(b)\overline{q}_t(d;\Gamma_t)} \quad (2)$$

$$f(\gamma) = 1 - f(C) = \frac{1}{1 + \sum_{t=1}^{5} \overline{q}(c)\overline{q}_t(d;\Gamma_t)} \quad (3)$$

$$f(\delta) = 1; f(D) = 0 \quad (4)$$

Equations (1)–(4): Mathematical expressions of the probability of disease presence for each type of URCOTEBS [7–10].

$$\lambda = \frac{Q_{u_i}^{\mathrm{nb}}}{Q_{u_i}^{a}} = \frac{\alpha A_a^D}{A_{\mathrm{nb}}^D} = 7.87 \quad (5)$$

$$\theta = \frac{V_\varepsilon^a}{V_\varepsilon^{\mathrm{nb}}} = 6.02 \quad (6)$$

$$V_\varepsilon = \frac{m_{\Omega_0}(q+3)}{4W\rho_\Omega} \quad (7)$$

Equations (5)–(7): The relations upon which the mathematical expressions of disease presence probability are founded for each type of URCOTEBS. These relations are the consequence of mathematical modeling (thermodynamic and geometric–statistical) of the structures and functions of the middle ear [1,2].

Table 9. Decoding stochastic information within any **URCOTEBS**, starting from the moment it was classified within one of the four conformations relative to the Disease Status of ME Mucosa.

Disease Status of D-Organ (Disease of the Middle-Ear Mucosa)								
URCOTEBS Generic Symbol	URCOTEBS Disease State Symbol	Initial Exclusive Random Event Index	Index Eveniment Aleator Exclusiv Ulterior	Initial White Balls (Mean Weight of Healthy D-Cells) [%]	Subsequent White Balls (Mean Weight of Healthy D-Cells) [%]	Initial Black Balls (Mean Weight of Diseased D-Cells) [%]	Subsequent Black Balls (Mean Weight of Diseased D-Cells) [%]	Probability
1	2	3	4	5	6	7	8	9
a	α	0–5	14–25	95	61	5	39	0.60%
b	β	6–10	14–25	84	61	16	39	1.40%
c	γ	11–13	14–25	76	61	24	39	7.90%
d	δ	14–25	no	61	no	39	no	100%

Table 10. Decoding stochastic information within any **URCOTEBS**, starting from the moment it was classified within one of the four conformations relative to the Health Status of ME Mucosa.

Health status of D-Organ (Health of the Middle-ear Mucosa)					
URCOTEBS Generic symbol	URCOTEBS Health State Symbol	Initial Exclusive Random Event Index	Initial White Balls (Mean Weight of Healthy D-Cells) [%]	Initial Black Balls (Mean Weight of Diseased D-Cells) [%]	Probability
1	2	3	4	5	6
a	A	0–5	95	5	99.40%
b	B	6–10	84	16	98.60%
c	C	11–13	76	24	92.10%

Due to the anatomic complexity, several mathematical models (static and dynamic behaviors) have been used to study ME function, all of them based on deterministic approaches [34–38].

Lumped element models, distributed parameter models, multi-body systems, and finite element models have been used to assess the function of the healthy ME [39,40], to evaluate pathological conditions [41], and also to verify results of ossicular chain reconstruction [42,43]. Because the high degree of variability among individuals leads to large discrepancies among samples, the ME behaves as an uncertain system, which means that stochastic models become very useful [34,44]. Modeling a dynamical system with uncertainties has been presented in the literature [45] and employs two groups of methods: parametric and nonparametric. Parametric approaches associate the uncertainties with the physical parameters of the system, and each parameter is treated as a random variable with a defined statistics, while nonparametric techniques deal with the uncertainties directly in the mathematical operators (matrices) of the system. The nonparametric probabilistic approach was proposed in 2000 by Soize et al., to consider the ME, and it was used to develop a stochastic model of the ME by Lobato et al., in 2022 [34,45,46]. It has also been applied in different contexts of uncertain structural dynamics, such as composite panels [34,47,48], automotive vehicles [34,49], drill-strings [34,50], and hermetic compressors [34,51]. Artificial intelligence (AI) shows promising results in the context of patient care, and several AI methods such as binary classification and logistic function could prove very useful for our future research [52]. Although the measuring gauge is currently manually drawn on the radiograph and is specific for each radiographic image, we envision the development of a software capable of directly applying the measuring gauge by using our theories. As previously shown, we use two equilateral triangles circumscribed to (1) the circular or oval transparency constituted by the almost total superimposing of the opacities of the two acoustic meati (external and internal); and (2) the opacity of the otic capsule and the tympanic bone. The second triangle circumscribes the endo-temporal cavity complex. We then calculate the ratio between the two surfaces to obtain the corresponding type of URCOTEBS.

Statistical calibration has also been used on geometry-based models such as the finite element (FE) model to calculate dynamic characteristics of the middle ear [39,53–56].

5. Conclusions

The unilateral radiographic image of the TB in Schuller's projection is a mirror that reflects the status of the ME mucosa, and **URCOTEBS** encodes the physiological state of the *D-Organ*. The present work gives the key to decoding this information into clinical language. The **URCOTEBS** matrix completely defines four types of unilateral radiographic conformations of the TB in Schuller's projection, which we termed, in ascending order of their projection areas (projection of their *Variable Geometry Peripheral Endo-temporal Bony Cavity Complex*) **URCOTEBS_d**, **URCOTEBS_c**, **URCOTEBS_b**, and **URCOTEBS_a**. The corresponding Greek letter designates the state of disease for each of these conformations: **URCOTEBS_δ**, **URCOTEBS_γ**, **URCOTEBS_β**, **URCOTEBS_α**, and the capital letters define their state of health: **URCOTEBS_D**, **URCOTEBS_C**, **URCOTEBS_B**, **URCOTEBS_A**.

URCOTEBS_d is the smallest unilateral radiographic conformation of the TB in Schuller's projection and is, by definition, a radiographic image of the state of disease of the *D-Organ*. The probability of disease in a **URCOTEBS_d** is 100%.

Starting from defining the *D-Organ*, we can now study all its essential aspects by using the unilateral radiographic image of the TB in Schuller's projection. This is achievable with stochastic analysis. These results are only a premise for future practical application. The discussion about the *D-Organ* as an *odd-numbered* organ is *necessary* but by no means *sufficient* both from a pathological and a clinical point of view, since both these perspectives require the study the *D-Organ* as a *pair*. This will be the subject of future work on **BI**lateral **R**adiographic **CO**nformation of the **TE**mporal **B**one in **S**chuller's projection (**BIRCOTEBS**). This radiographic system is very practical, cheap, readily available, and clinically usable as we defined in this paper.

A probabilistic reasoning is one that most people use to draw conclusions from incomplete or imperfect data. Even though measuring, reasoning, or communicating are possible without the use of probability theory, we consider it, as others have before us, indispensable and unavoidable in science. The material universe, including living creatures, is made up, as the name suggests, of matter. Some aspects of life will not be readily available for explanations employing current scientific techniques. It is not considered necessary nor possible to assume that living matter as analyzed by scientists could contain anything considered mysterious by a chemist or a physician. Thus, the fundamental laws of physical matter are nothing but the fundamental laws of any matter. Since the behavior of molecules, atoms, and elementary elements is analyzed with probabilistic theories by quantum physics, we must acknowledge the use of probabilistic estimations for the present topic as well.

Although famous scientists such as *Albert Einstein (1879–1955)* were unhappy with probabilistic foundations for science (*"God does not play dice"*), most members of the scientific community including such giants as *Niels Bohr (1885–1962)* or *Werner Heisenberg (1901–1976)* accept the modern rule or probability theory. If we understand that the most basic laws of elementary physical phenomena are probabilistic, we can acknowledge that probability theory is essential for many special scientific fields, including medicine.

Author Contributions: M.R., A.-C.N. and H.M. have contributed equally to this work and should, therefore, be considered first authors of this article. Conceptualization: M.R. and H.M.; methodology: M.R., H.M. and A.-C.N.; software: A.-I.M. and M.-A.S.; validation: M.R., A.-C.N. and H.M.; formal analysis: A.-I.M., H.M. and A.-C.N.; investigation: M.R., H.M. and A.-C.N.; resources: A.-I.M., M.-A.S. and H.M.; data curation: H.M., M.-A.S. and A.-I.M.; writing—original draft preparation: M.R. and H.M.; writing—review and editing: M.R. and A.-C.N.; visualization: M.R., H.M. and A.-C.N.; supervision: H.M. and M.R.; project administration: H.M., M.R. and A.-C.N. A.-I.M. and M.-A.S. have made substantial contributions to the acquisition, analysis, and interpretation of data and performed a final view of the article. A.-I.M. and M.-A.S. were also responsible of the data analysis and graphical representation of the results and performed a final view of the article. All authors have read and agreed to the published version of the manuscript.

Funding: This research received no external funding.

Institutional Review Board Statement: Not applicable.

Informed Consent Statement: Not applicable.

Data Availability Statement: Data are contained within the article.

Conflicts of Interest: The authors declare no conflict of interest.

References

1. Rădulescu, M.; Mocanu, H.; Nechifor, A.; Mocanu, A.I. Thermodynamic Model for Middle-Ear Ventilation-Defining The D-Organ by Comparison to The Eustachian Tube (Re-Interpretation of A Classic Experiment and Review of Literature). *Acta Tech. Napocensis Ser. Appl. Mat. Mech. Eng.* **2022**, *65*, 69–78.
2. Rădulescu, M.; Mocanu, A.-I.; Dascălu, I.T.; Schipor, M.-A.; Mocanu, H. Geometric–Statistical Model for Middle-Ear Anatomy and Ventilation. *Appl. Sci.* **2022**, *12*, 11287. [CrossRef]
3. Broussais, F.J.V. *Traité de Physiologie Appliquée à la Pathologie*, 1822 ed.; Hachette BnF: Paris, France, 2013.
4. Canhuilhem, G. *On the Normal and the Pathological*; Reidel Publishing Company: Dordrecht, The Netherlands, 1978. [CrossRef]
5. Shelshakova, N.N. "Norm" and "Pathology": A Comparative Analysis of Concepts. *Nat. Volatiles Essent. Oils* **2021**, *8*, 7377–7383.
6. Rostand, J. *Claude Bernard–Morceaux Choisis, Cap I, Leçons de Physiologie Experimentale appliquée à la Medicine, Sur le Rôle de L'anatomie Dans la Découverte des Fonctions*, Septième ed.; Gallimard: Paris, France, 1938; pp. 37–48. (In French)
7. Văduva, I. On a Particular Lifetime Distribution. *Rev. Air Force Acad.* **2017**, *2*, 5–14. [CrossRef]
8. Fricker, C.; Guillemin, F.; Robert, P. An Identification Problem in an Urn and Ball Model with Heavy Tailed Distributions. *Probab. Eng. Informational Sci.* **2010**, *24*, 77–97. [CrossRef]
9. Busbee, B. Uses of the Hypergeometric Distribution for Determining Survival or Complete Representation of Subpopulations in Sequential Sampling. Master's Thesis. Electronic Theses and Dissertations, Stephen F Austin State University, Nacogdoches, TX, USA, 2017; p. 118. Available online: https://scholarworks.sfasu.edu/etds/118 (accessed on 23 November 2023).
10. Ghitany, M.E.; Al-Mutairi, D.K.; Nadarajah, S. Zero-truncated Poisson–Lindley distribution and its application. *Math. Comput. Simul.* **2008**, *79*, 279–287. [CrossRef]
11. Neagu, A.C.; Budișteanu, M.; Gheorghe, D.C.; Mocanu, A.I.; Mocanu, H. Rare Gene Mutations in Romanian Hypoacusis Patients: Case Series and a Review of the Literature. *Medicina* **2022**, *58*, 1252. [CrossRef] [PubMed]

12. Mocanu, H. The role of perinatal hearing screening in the normal development of the infant's language. In *Debating Globalization. Identity, Nation and Dialogue*, 4th ed.; Boldea, I., Sigmirean, C., Eds.; Arhipeleag XXI Press: Tirgu Mures, Romania, 2017; pp. 562–569.
13. Mocanu, H. The economic impact of early diagnosis of congenital hearing loss. In *Debating Globalization. Identity, Nation and Dialogue*, 4th ed.; Boldea, I., Sigmirean, C., Eds.; Arhipeleag XXI Press: Tirgu Mures, Romania, 2017; pp. 556–561.
14. Swarts, J.D.; Doyle, B.M.C.; Doyle, W.J. Surface Area-Volume Relationships for the Mastoid Air Cell System in Adult Humans. *J. Laryngol. Otol.* **2011**, *125*, 580–584. [CrossRef]
15. Sade, J.; Fuchs, C. A comparison of mastoid pneumatization in adults and children with cholesteatoma. *Eur. Arch. Otorhinolaryngol.* **1994**, *251*, 191–195. [CrossRef]
16. Sade, J.; Fuchs, C. Secretory Otitis Media in Adults: I. The Role of Mastoid Pneumatization as a Risk Factor. *Ann. Otol. Rhinol. Laryngol.* **1996**, *105*, 643–647. [CrossRef]
17. Sadé, J.; Fuchs, C. Secretory Otitis Media in Adults: II. The Role of Mastoid Pneumatization as a Prognostic Factor. *Ann. Otol. Rhinol. Laryngol.* **1997**, *106*, 37–40. [CrossRef]
18. Lesinskas, E. Factors affecting the results of nonsurgical treatment of secretory otitis media in adults. *Auris Nasus Larynx* **2003**, *30*, 7–14. [CrossRef] [PubMed]
19. Valtonen, H.J.; Dietz, A.; Qvarnberg, Y.H.; Nuutinen, J. Development of Mastoid Air Cell System in Children Treated with Ventilation Tubes for Early-Onset Otitis Media: A Prospective Radiographic 5-Year Follow-Up Study. *Laryngoscope* **2005**, *115*, 268–273. [CrossRef] [PubMed]
20. Doyle, W.J. The mastoid as a functional rate-limiter of middle ear pressure change. *Int. J. Pediatr. Otorhinolaryngol.* **2007**, *71*, 393–402. [CrossRef] [PubMed]
21. Cohen, D.; Raveh, D.; Peleg, U.; Nazarian, Y.; Perez, R. Ventilation and clearance of the middle ear. *J. Laryngol. Otol.* **2009**, *123*, 1314–1320. [CrossRef] [PubMed]
22. Byun, S.W.; Lee, S.-S.; Park, J.Y.; Yoo, J.H. Normal Mastoid Air Cell System Geometry: Has Surface Area Been Overestimated? *Clin. Exp. Otorhinolaryngol.* **2016**, *9*, 27–32. [CrossRef] [PubMed]
23. Park, M.S.; Yoo, S.H.; Lee, D.H. Measurement of surface area in human mastoid air cell system. *J. Laryngol. Otol.* **2000**, *114*, 93–96. [CrossRef]
24. Koç, A.; Ekinci, G.; Bilgili, A.M.; Akpinar, I.N.; Yakut, H.; Han, T. Evaluation of the mastoid air cell system by high resolution computed tomography: Three-dimensional multiplanar volume rendering technique. *J. Laryngol. Otol.* **2003**, *117*, 595–598. [CrossRef]
25. Palva, T.; Palva, A. Size of the Human Mastoid System. *Acta. Otolaryngol. Stockh.* **1966**, *62*, 237–251. [CrossRef]
26. Csakanyi, Z.; Katona, G.; Josvai, E.; Mohos, F.; Sziklai, I. Volume and Surface of the Mastoid Cell System in Otitis Media with Effusion in Children: A Case-Control Study by Three-Dimensional Reconstruction of Computed Tomographic Images. *Otol. Neurotol.* **2011**, *32*, 64–70. [CrossRef]
27. Cinamon, U. The growth rate and size of the mastoid air cell system and mastoid bone: A review and reference. *Eur. Arch. Oto-Rhino-Laryngol.* **2009**, *266*, 781–786. [CrossRef]
28. Hug, J.E.; Pfaltz, C.R. Temporal bone pneumatization. A planimetric study. *Arch. Otorhinolaryngol.* **1981**, *233*, 145–156. [CrossRef]
29. Flisberg, K.; Zsigmond, M. The Size of the Mastoid Air Cell System: *Planimetry—Direct Volume Determination. Acta Oto-Laryngol.* **1965**, *60*, 23–29. [CrossRef]
30. Todd, N.W.; Pitts, R.B.; Braun, I.F.; Heindel, H. Mastoid size determined with lateral radiographs and computerized tomography. *Acta Otolaryngol.* **1987**, *103*, 226–231. [CrossRef] [PubMed]
31. Vrabec, J.T.; Champion, S.W.; Gomez, J.D.; Johnson, R.F., Jr.; Chaljub, G. 3D CT imaging method for measuring temporal bone aeration. *Acta. Otolaryngol.* **2002**, *122*, 831–835. [CrossRef] [PubMed]
32. Lee, D.H.; Jun, B.C.; Kim, D.G.; Jung, M.K.; Yeo, S.W. Volume variation of mastoid pneumatization in different age groups: A study by three-dimensional reconstruction based on computed tomography images. *Surg. Radiol. Anat.* **2005**, *27*, 37–42. [CrossRef]
33. Colhoun, E.N.; O'Neill, G.; Francis, K.R.; Hayward, C. A comparison between area and volume measurements of the mastoid air spaces in normal temporal bones. *Clin. Otolaryngol. Allied Sci.* **1988**, *13*, 59–63. [CrossRef] [PubMed]
34. Lobato, L.C.; Paul, S.; Cordioli, J.A.; Ritto, T.G. Stochastic model of the human middle ear using a nonparametric probabilistic approach. *J. Acoust. Soc. Am.* **2022**, *151*, 2055–2065. [CrossRef]
35. De Greef, D.; Buytaert, J.A.; Aerts, J.R.; Van Hoorebeke, L.; Dierick, M.; Dirckx, J. Details of human middle ear morphology based on micro-CT imaging of phosphotungstic acid stained samples. *J. Morphol.* **2015**, *276*, 1025–1046. [CrossRef]
36. Rosowski, J.; Chien, W.; Ravicz, M.; Merchant, S. Testing a Method for Quantifying the Output of Implantable Middle Ear Hearing Devices. *Audiol. Neurotol.* **2007**, *12*, 265–276. [CrossRef]
37. Funnell, W.R.J.; Maftoon, N.; Decraemer, W.F. *Modeling of Middle Ear Mechanics*; Springer: New York, NY, USA, 2013; pp. 171–210.
38. Paolis, A.D.; Bikson, M.; Nelson, J.T.; de Ru, J.A.; Packer, M.; Cardoso, L. Analytical and numerical modeling of the hearing system: Advances towards the assessment of hearing damage. *Hear. Res.* **2017**, *349*, 111–128. [CrossRef]
39. Koike, T.; Wada, H.; Kobayashi, T. Modeling of the human middle ear using the finite-element method. *J. Acoust. Soc. Am.* **2002**, *111*, 1306–1317. [CrossRef]

40. Lobato, L.; Paul, S.; Cordioli, J. Statistical analysis of the human middle ear mechanical properties. *J. Acoust. Soc. Am.* **2022**, *151*, 2043–2054. [CrossRef]
41. Gan, R.Z.; Cheng, T.; Dai, C.; Yang, F.; Wood, M.W. Finite element modeling of sound transmission with perforations of tympanic membrane. *J. Acoust. Soc. Am.* **2009**, *126*, 243–253. [CrossRef] [PubMed]
42. Neudert, M.; Bornitz, M.; Mocanu, H.; Lasurashvili, N.; Beleites, T.; Offergeld, C.; Zahnert, T. Feasibility Study of a Mechanical Real-Time Feedback System for Optimizing the Sound Transfer in the Reconstructed Middle Ear. *Otol. Neurotol* **2018**, *39*, e907–e920. [CrossRef] [PubMed]
43. Mocanu, H.; Bornitz, M.; Lasurashvili, N.; Zahnert, T. Evaluation of Vibrant® Soundbridge™ positioning and results with laser doppler vibrometry and the finite element model. *Exp. Ther. Med.* **2021**, *21*, 262. [CrossRef] [PubMed]
44. Ebrahimian, A.; Maftoon, N. Stochastic finite element modelling of human middle-ear. In Proceedings of the 44nd Conference of The Canadian Medical and Biological Engineering, Virtual Conference, 11–13 May 2021; Volume 44.
45. Daouk, S.; Louf, F.; Dorival, O.; Champaney, L.; Audebert, S. Uncertainties in structural dynamics: Overview and comparative analysis of methods. *Mech. Ind.* **2015**, *16*, 404. [CrossRef]
46. Soize, C. A nonparametric model of random uncertainties for reduced matrix models in structural dynamics. *Probabilistic Eng. Mech.* **2000**, *15*, 277–294. [CrossRef]
47. Chen, C.; Duhamel, D.; Soize, C. Probabilistic approach for model and data uncertainties and its experimental identification in structural dynamics: Case of composite sandwich panels. *J. Sound Vib.* **2006**, *294*, 64–81. [CrossRef]
48. Petrescu, S.M.S.; Țuculina, M.J.; Popa, D.L.; Duță, A.; Salan, A.I.; Voinea Georgescu, R.; Diaconu, O.A.; Turcu, A.A.; Mocanu, H.; Nicola, A.G.; et al. Modeling and Simulating an Orthodontic System Using Virtual Methods. *Diagnostics* **2022**, *12*, 1296. [CrossRef] [PubMed]
49. Durand, J.-F.; Soize, C.; Gagliardini, L. Structural-acoustic modeling of automotive vehicles in presence of uncertainties and experimental identification and validation. *J. Acoust. Soc. Am.* **2008**, *124*, 1513–1525. [CrossRef] [PubMed]
50. Ritto, T.; Soize, C.; Sampaio, R. Robust optimization of the rate of penetration of a drill-string using a stochastic nonlinear dynamical model. *Comput. Mech.* **2010**, *45*, 415–427. [CrossRef]
51. Fontanela, F.; Silva, O.; Lenzi, A.; Ritto, T. Development of a stochastic dynamical model for hermetic compressor components with experimental investigation. *Mech. Syst. Sign. Process.* **2016**, *76*, 712–728. [CrossRef]
52. Kufel, J.; Bargieł-Łączek, K.; Kocot, S.; Ko'zlik, M.; Bartnikowska, W.; Janik, M.; Czogalik, Ł.; Dudek, P.; Magiera, M.; Lis, A.; et al. What Is Machine Learning, Artificial Neural Networks and Deep Learning? —Examples of Practical Applications in Medicine. *Diagnostics* **2023**, *13*, 2582. [CrossRef] [PubMed]
53. Lee, D.; Ahn, T.-S. Statistical calibration of a finite element model for human middle ear. *J. Mech. Sci. Technol.* **2015**, *29*, 2803–2815. [CrossRef]
54. Volandri, G.; Di Puccio, F.; Forte, P.; Manetti, S. Model-oriented review and multi-body simulation of the ossicular chain of the human middle ear. *Med. Eng. Phys.* **2012**, *34*, 1339–1355. [CrossRef]
55. Gan, R.; Reeves, B.; Wang, X. Modeling of sound transmission from ear canal to cochlea. *Ann. Biomed. Eng.* **2007**, *35*, 2180–2195. [CrossRef]
56. Gan, R.Z.; Sun, Q.; Feng, B.; Wood, M.W. Acoustic structural coupled finite element analysis for sound transmission in human ear–Pressure distributions. *Med. Eng. Phys.* **2006**, *28*, 395–404. [CrossRef]

applied sciences

MDPI

Article

Designing a Synthetic 3D-Printed Knee Cartilage: FEA Model, Micro-Structure and Mechanical Characteristics

Gianmarco Dolino [1,†], Damiano Coato [1,†], Riccardo Forni [1,2,*], Gabriele Boretti [1], Federica Kiyomi Ciliberti [1] and Paolo Gargiulo [1,3]

1 Institute of Biomedical and Neural Engineering, Reykjavik University, 102 Reykjavík, Iceland; dolinogianmarco@gmail.com (G.D.); coatodamiano@gmail.com (D.C.); gabrieleb@ru.is (G.B.); federica21@ru.is (F.K.C.); paolo@ru.is (P.G.)
2 Department of Electric, Electronic and Information Engineering Guglielmo Marconi, Bologna University, 40136 Bologna, Italy
3 Department of Science, Landspitali Huniversity Hospital, 105 Reykjavik, Iceland
* Correspondence: riccardo21@ru.is
† These authors contributed equally to this work.

Abstract: Articular cartilage morphology and composition are essential factors in joint biomechanics, and their alteration is a crucial aspect of osteoarthritis (OA), a prevalent disease that causes pain and functional loss. This research focuses on developing patient-specific synthetic cartilage using innovative Digital Anatomy polymers. The objectives include investigating the morphology, characterizing the mechanical properties, and replicating the architecture of natural cartilage. This approach offers potential alternatives to traditional manufacturing methods and reduces the need for expensive in vivo experiments. Finite Element Analysis (FEA) validates a novel patient-specific measurement setup. It provides insights into the role of morphology in the distribution of stress and strain within cartilage. CAD design is also utilized to create standardized fiber-reinforced samples that mimic the layered micro-architecture of natural cartilage, allowing for the study of their contribution to the overall mechanical properties. The results demonstrate that 3D-printed polymers can effectively replicate the elastic properties of cartilage. The proposed patient-specific simulator produces reliable results, which have been validated through FEM analysis. While the recreated microstructure closely resembles biological cartilage samples, the elastic properties are slightly underestimated. In conclusion, designing an in silico knee joint is a feasible approach that offers numerous advantages for further development. The Young's modulus values of our synthetic cartilage modules range from 2.43 MPa to 7.24 MPa, within the range reported in the literature. Moreover, Young´s modulus at the micro level shows the differences between surface 1.74 MPa and internal substrate 1.83 MPa depending on the fiber orientation. Finally, our model proves to be mechanically and morphologically accurate at both the macro and micro levels.

Keywords: knee; cartilage; 3D printing; segmentation; finite element analysis

1. Introduction

Articular cartilage is a complex anisotropic tissue composed of a high percentage of water and biological components, such as collagen and glycosaminoglycans (GAGs), and plays a pivotal role in providing a cushioning effect, reducing friction between bones, and ensuring seamless articulation. The mechanical performance of native cartilage is mainly contributed by collagen fibers and their orientation throughout the tissue. The cartilage can be divided into four zones based on depth: superficial zone (10–20% of total thickness); middle zone (40–60%); deep zone (20–30%); and calcified zone [1,2]. These zones differ in composition, with variations in GAGs and collagen content, collagen fiber orientation, and cell density. The transition between the deep and calcified zones is known as the tidemark, where the cartilage tissue merges into the subchondral bone.

Osteoarthritis (OA) is the most common form of cartilage and joint disease globally [3]. It is a prevalent cause of pain, functional loss, and adult disability in Western countries [4].

OA is characterized by the gradual erosion and degeneration of the articular cartilage within the joint, which serves as a critical component in facilitating smooth joint movement. However, in OA, this protective cartilage undergoes significant changes. As the disease progresses, the cartilage begins to wear away, leading to an uneven distribution of mechanical forces within the joint. This imbalance triggers the generation of fibrocartilage, an inferior type of tissue that lacks the resilience and functionality of healthy cartilage. Concurrently, catabolic reactions ensue, further contributing to the breakdown of the cartilaginous matrix. The non-physiological mechanical stimulation of chondrocytes, the specialized cells within the cartilage responsible for maintaining its structure, adds another layer of complexity to the OA pathophysiology. These chondrocytes, instead of promoting cartilage repair, become active participants in the degenerative process [5].

Currently, the gold standards for diagnosing knee OA include clinical evaluation, medical history, and imaging techniques. Physical examination and medical history help assess symptoms, joint deformities, and previous joint-related conditions. Imaging techniques such as X-ray, Magnetic Resonance Imaging (MRI), and Computed Tomography (CT) provide valuable information on the degree and location of joint damage, confirming the diagnosis and assessing the extent of damage [6].

Unfortunately, there is currently no disease-modifying therapy available for OA, and medications aimed at treating its symptoms are generally not very successful. Therefore, the focus in cartilage treatments is to develop methods with consistent regenerative potential [7]. Animal models are crucial for bridging the gap between in vitro experiments and human clinical studies when introducing these treatments into clinical practice. Choosing an appropriate animal model involves considering multiple factors, including research funding, costs of animal purchase and housing, and the suitability of each model for specific cartilage research projects [8].

Amid increasing awareness about the importance of reducing animal experiments, cutting costs, and accelerating development timelines, our study aims to design and create synthetic cartilage tissue using computational modeling and 3D-printing technology. Three-dimensional printing, also known as additive manufacturing, involves the layer-by-layer deposition of materials to create three-dimensional objects. It is widely used in various industries for producing prototypes, customized products, and complex geometries [9]. In healthcare, 3D printing is utilized for creating patient-specific implants [10], prosthetics, and anatomical models for surgical planning [11]. While previous 3D-printed anatomical models have achieved similarities to human tissue in shape, they have not fully replicated mechanical responsiveness [12]. Digital Anatomy Printing is an advanced form of 3D printing specifically designed for the medical field. It involves the use of specialized materials that mimic the properties of human tissues, allowing for the creation of highly detailed anatomical models [13]. Digital Anatomy Printing bridges the gap between traditional 3D printing and BioPrinting by simulating the look and feel of human tissues without the ability to create functional living tissues [14]. On the other hand, BioPrinting stands out for its ability to create functional living tissues and organs, ushering in a new era of possibilities in the fields of medicine and biotechnology [15]. However, the dimension is still limited due to the absence of vascularization [16]; the mechanical properties of these functionalized tissues have not reached yet those of their native counterparts [17].

This proposed methodology not only provides a promising alternative to traditional manufacturing methods but also enables the use of Finite Element Analysis (FEA) to assess the mechanical characteristics of materials in a simulated environment, reducing the need for expensive and complex in vivo experiments [12,18]. In this article, we will explore the process of developing patient-specific synthetic cartilage tissue, with a specific focus on mechanical properties, morphology, and microstructure.

The main goals of the study are (1) the morphological investigation and characterization of the knee joint in healthy and osteoarthritic patients with the segmentation of medical images through FEM and mechanical tests; (2) the mechanical characterization of 3D-printed innovative Digital Anatomy polymers capable of matching the behavior of cartilage native tissue; (3) mimicking the cartilage morphology by taking into account the micro-structure of the tissue and thus incorporating the fibers inside the samples and assessing their contribution to the mechanical response.

2. Materials and Methods

2.1. Clinical Data

Patients diagnosed with Degenerative OA (D) and healthy Control subjects (C) were recruited within the EU project RESTORE (https://restore-project.ru.is, accessed 1 July 2022). A further explanation of the acquisition protocol and the facilities utilized is described in a previous work [19].

Within the aim of relating material properties and morphological changes due to OA, one young healthy subject and one elder with clinically evaluated cartilage degeneration were chosen. The subjects description is as follows:

- Subject 1: group = D, age = 68, sex = M, knee = left, BMI = 27.47;
- Subject 2: group = C, age = 26, sex = M, knee = left, BMI = 22.28.

The medical imaging processing utilized is based on a workflow specifically developed for knee cartilages and bones [20].

From the various anatomical components, only femoral cartilage, femur and tibia data were extracted. The choice was made in order to put the focus on the cartilage itself while avoiding unnecessary computational costs and material waste as it will be further explained in the following chapters.

2.2. Meshes Generation for FEM Analysis and 3D Printing

Geometries of the femur, tibia and cartilage were imported on the CAD software 3-Matic 17.0 (Materialise, Belgium) to generate the needed superficial and volumetric mesh for FEM and 3D-printing steps. Since the focus of the analysis is on cartilage, the bones were meshed coarsely to reduce the computational requirements, while a finer mesh was used for the femoral cartilage. Specifically, a 2D triangular mesh with a size of 4 mm was used for most of the bone surface, while a 2 mm triangular mesh was used for the cartilage. To achieve a better load distribution at the transition from bone to cartilage, called the chondral zone, the bone mesh was refined towards the contact zones with the cartilage, ultimately reaching 2 mm. Moreover, a 1 mm mesh was employed to finely mesh the cartilage regions containing geometric singularities (holes caused by chondral defects).

The volume mesh was generated using quadratic tetrahedral elements (TET10) for femoral cartilage, while for the femur and tibia, as the bone components were of the least focus, were used linear tetrahedral elements (TET4) [21]. In the Control model, the average 3D element mesh size measured 0.81 mm, whereas in the Degenerative model, it was 0.79 mm.

Anatomical supports for mechanical testing were also designed to realize a physical model of the knee joint to compress cartilages with standard universal testing machines. A 80 × 80 mm square was created with the intent of providing a surface area sizable enough to guarantee adequate compression force distribution. A 25 × 25 × 10 mm rectangle was created to enable accurate closure between the testing machine's grips and suitable positioning of the square support on the end of the grips. The superficial and volume meshes generated for the anatomical supports were the same as mentioned above.

2.3. Design of Cartilage Tissue Micro-Structure

To replicate the micro-structure of femoral cartilage and the orientation of type II collagen fibers in different layers, three fiber-reinforced materials were designed. Collagen-like fibers were realized and integrated into ASTM D695 and D638 standardized samples (Figure 1) on the open-source 3D computer graphics software Blender 4.0. The minimum fiber diameter was set to 60 µm due to the printer resolution [13]. The length was accordingly scaled up to maintain the aspect ratio of the single fiber.

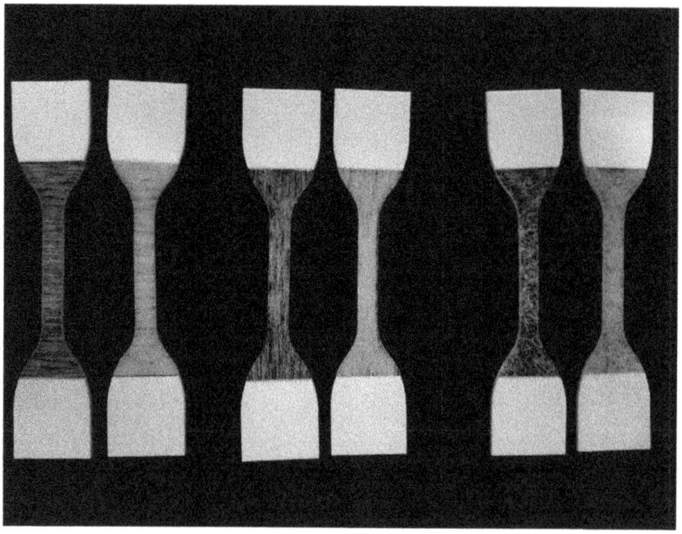

Figure 1. Standardized samples (ASTM D638) with different fiber orientations and concentrations. From left to right, respectively, we have the following: Superficial (1% and 5%), Deep (1% and 5%), and Middle (1% and 5%).

The superficial zone, in the specimens, was represented by fibers oriented tangentially to the load axis. Beneath the superficial zone is the middle zone: fibers representative of this layer were randomly oriented, with a tendency for a 45° angle. To mimic the deep zone, fibers were oriented in parallel with respect to the long axis of the specimens. Our design did not include the calcified zone due to its mixed composition, which positions it between cartilage and bone mechanical response.

The fibers were designed with a standardized length due to a lack of uniformity in the measurement results of the biological samples [22,23]. On the other hand, there are several proofs that the fibril diameter changes along with the cartilage depth [24]. From the superficial to the deep layer, the diameter increases due to the fibers aggregate. As can be seen in Table 1, we designed three different fiber types mimicking diameter changes.

Collagen generally accounts for a maximum of a 15–20% of the total wet weight of cartilage tissue [24,25]. To study the effect of fibers integration on the composite's mechanical properties, 1% and 5% volume contents were chosen to contain the computational effort required to handle and print complex structures (Figure 1).

Table 1. Dimensions of designed fibers escalating through consecutive layers, expressed in mm.

Layer	Length	Diameter
Superficial	6	0.06
Middle	6	0.12
Deep	6	0.18

2.4. Synthetic Materials and 3D Printing

The blends of materials used to replicate the characteristics of native cartilage tissue were designed using the GrabCAD Digital Anatomy Creator 1.73 (GrabCAD-Stratasys, Cambridge, MA, USA) software. The objective was to create a composition that closely mimicked the mechanical properties of natural cartilage.

For this purpose, a combination of three materials was utilized: *BoneMatrix*, *Agilus*30, and *GelMatrix*. *BoneMatrix* is an elasto-plastic material with Young's modulus ranging from 2000 to 3000 MPa and was chosen to incorporate the desired stiffness into the synthetic cartilage. *Agilus*30, on the other hand, is a flexible rubber-like material with a shore hardness of 30A, which was essential for replicating the flexibility of natural cartilage. Lastly, *GelMatrix* is a polymeric, uncompressible gel specifically designed to mimic the compressibility of real cartilage.

Two different composites were created using these materials. The first composite called *Soft Cartilage*, consisted of 70% flexible material, 15% gel polymer, and 15% stiff polymer. The second composite, named *Medium Cartilage*, had a composition of 85% flexible material and 15% stiff polymer. These variations were introduced to study the mechanical properties of the synthetic cartilage under different degrees of flexibility and stiffness.

To fabricate these synthetic materials, the J850 Digital Anatomy Printer (Stratasys, Eden Prairie, MN, USA), a cutting-edge 3D printer specifically designed for anatomical models, was employed. Additionally, the Objet260 Connex 3 (Stratasys, Eden Prairie, MN, USA) was used to print anatomically shaped supports for the subsequent mechanical tests. These supports were printed using the rigid *VeroWhite* material, ensuring stable and accurate testing conditions (Figure 2).

In cases where fiber reinforcement was required, a blend of 50% *Agilus*30 and *TissueMatrix* material was used. The fibers themselves were printed using the rigid *VeroWhite* material, enabling the incorporation of reinforced structural elements in the synthetic cartilage samples.

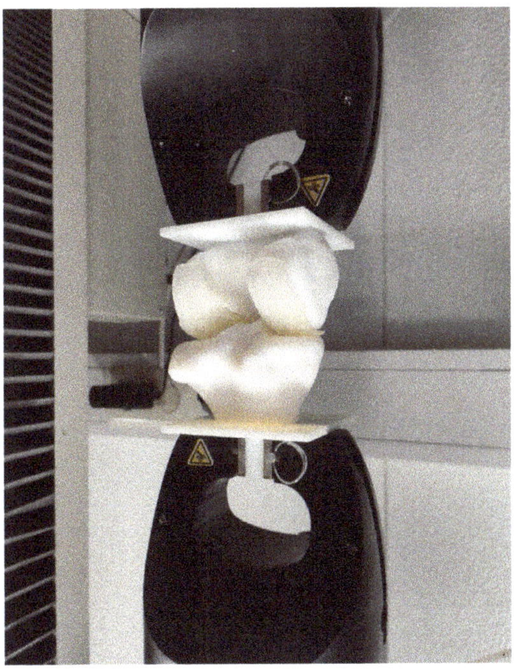

Figure 2. Three-dimensionally printed functional model of the knee joint (femur, tibia and femoral cartilage) during the mechanical evaluation.

2.5. Finite Element Method Analysis

On ANSYS 2022 (Canonsburg, PA, USA), models of the bones (with and without supports), and femoral cartilage from each subject were imported to understand the influence of OA defects onto the cartilage mechanical properties through computational analysis. Here, the different parts were assembled into a single model, and the material properties, loads and boundary conditions were assigned. Five different materials were chosen during the simulation: *Cartilage Literature*, *Bone Literature*, *Printed Cartilage Soft*, *Printed Cartilage Medium* and *Printed Bone*.

From a biomechanical standpoint, articular cartilage can be effectively represented as an elastic isotropic material during dynamic loading [26]. This is because the loading time of interest, corresponding to that of a fully extended leg touching the ground, is much shorter than the viscoelastic time constant of cartilage, which is 1500 s [27]. This approximation proves accurate for short-term cartilage response as demonstrated by Donzelli et al. [28], who proved that there are no significant changes in the cartilage contact response shortly after loading. Cartilage behaves as an incompressible material in its short-term (instantaneous) response ($\nu \approx 0.5$), while it reflects characteristics of a compressible material in its long-term response ($\nu \approx 0$) [29]. Therefore, in order to capture the intermediate behavior observed during physiological walking, which falls between the instantaneous and long-term responses to functional loads, a Young's modulus of 5 MPa and a Poisson's ratio of 0.45 were chosen to represent the cartilage behavior [30–32].

We assessed the impact of bone properties on cartilage with a FEM analysis. We assigned the materials *Printed Bone* and *Bone Literature* to the bones and then evaluated the results. Despite the substantial difference in stiffness values between the two materials, the observed variations in stress and strain within the femoral cartilage were close to 1%. Therefore, since the influence of bones properties on cartilage was minimal, we decided to model the bones as an isotropic linear elastic material with a Young's modulus of 17,000 MPa and a Poisson's ratio of 0.31 [33] in order to avoid unnecessary computational costs as confirmed in the literature [31,34]. Elastic properties of the designed materials are reported in Table 2.

Table 2. Material properties used in the simulation; Printed Cartilage modules were calculated as stated in Section 2.6.

Name	Young's Modulus (MPa)
Cartilage Literature	5.00
Printed Cartilage Soft	2.43
Printed Cartilage Medium	7.24
Bone Literature	17,000.00
Printed Bone	2500.00

The connections between femur–cartilage and cartilage–tibia were defined as bonded [35] due to a partially mineralized zone between that secure cortical bone and the articular cartilage by anchoring the collagen fibrils of the deep zone to the subchondral bone [1]. Due to the simplification applied to the model, by excluding menisci and tibial cartilages, a pinball region was set to compensate for the gap between the tibia and femoral cartilage. The asymmetric behavior and the augmented Lagrange formulation were used for handling contacts. In the analyses, flexo-extension and varus-valgus rotations were constrained for the femur in order to analyze the knee joint in full extension. The tibia was constrained on the lower surface with a fixed support. A compressive force of 1150 N was applied perpendicularly to the upper surface of the femur, which matches the force acting during a gait cycle for a full extension position [36] (Figure 3, Left). The values of Equivalent Elastic Strain [mm/mm] and Equivalent (von Mises) Stress [MPa] for the femoral cartilage were exported, and the stress–strain graphs were created to analyze the joint behavior with different material properties.

The first FEM analyses on both subjects were carried out by assigning the *Soft* and *Medium* materials to the femoral cartilages and the *Printed Bone* (*VeroWhite*) material to the bones, using their native geometry. Subsequently, to compare and validate the obtained results, a further analysis was conducted by assigning to the models the literature values of cartilage and bone tissues. After that, the same analysis was performed on the models designed for 3D printing to evaluate the effect of the supports and the compression mechanism. Bone material's properties were changed to *Printed Bone*, and thus the cartilage's ones (*Soft* and *Medium*). In the real setting, the grip is clamped on the sides but, after a simulation, it turned out to be equivalent to applying the 1150 N load on the upper surface of the grip in the computational environment (Figure 3, Right).

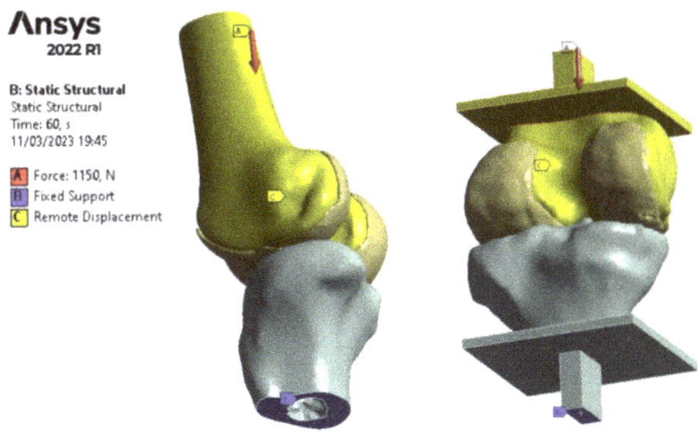

Figure 3. Boundary conditions and loading configuration utilized during the FEM analysis.

2.6. Mechanical Tests

The mechanical characterization of the printed specimens and anatomical models was carried out with the MTS Insight™ (Electromechanical-10 kN Standard Length) testing system at Tæknisetur (Reykjavik, Iceland).

The mechanical properties of the two custom materials were evaluated using a total of 20 standardized samples. Specifically, ten cylindrical samples were printed to meet the dimensional requirements outlined by the ASTM D695 compression protocol, five using the *Soft* material and five using the *Medium* material. Additionally, ten dog-bone specimens were printed for the ASTM D638 tensile test, with an equal split of five for each material type. The Young's modulus of each polymeric blend was computed, as an average, from the stress–strain charts in the linear region. This was done by following the corresponding ASTM guidelines for the calculation of elastic parameters.

For patient-specific testing (Figure 2), a total of 20 synthetic cartilages were realized, 10 for healthy subjects and 10 for degenerative conditions. Each patient had five cartilages printed with the *Soft* material and five with the *Medium* material. Before the test, every cartilage was measured with a digital caliber in six different points: two for the width and four for the thickness. Both the femur and tibia were fixed, and the jaws were positioned to keep the cartilage perfectly in place, as in a physiological condition. To secure the positioning, a pre-load of 650 N was applied to the cartilage. This choice also has physiological justification since the loads acting on the knee joint (for a leg in full extension position) during the standing phase of the gait cycle range from 700 N to 1150 N on average [31]. The compressive load was increased at a uniform rate of 1.3 mm/min along the longitudinal direction until a 20% strain of the total average thickness was reached. After that, the cartilage underwent the same measurements described above. This procedure was carried out for both subjects.

A total of 36 fiber-reinforced samples were realized and tested under uni-axial compression and tension to extract the Young's modulus. The results were then compared to assess the contribution of fibers dimension and orientation. Half of these were printed with 1% fiber content, while the remaining had 5% fiber content. Within each fiber volume, 18 samples were created—3 *Superficial*, 3 *Middle*, and 3 *Deep*—both for the tensile and compression tests. The zones of the dog-bone samples that are clamped during the measurements were printed with a reinforced material (*VeroWhite*) to ensure a solid grip as can be seen in Figure 1. Several tests were conducted to exclude the possibility for the samples to break at the transition zone.

3. Results

3.1. Images Post-Processing and FEA Results

From the image analysis, several characteristics were extracted, including the cartilage volume. Specifically, for the femoral cartilage, subject 1 (D) presents volume = 23,682.14 mm^3, and subject 2 (C) presents volume = 16,401.17 mm^3.

The FEA was used to validate what was measured with the synthetic knee because it allows for a controlled environment, where the geometries are aligned and changes due to OA can show the effect of the degeneration. The results shown in Figure 4 represent the strain distribution on the healthy (upper row) and osteoarthritic (lower row) morphologies taken into account for the computational simulation. *Medium* material (left column) exhibits the lowest deformation, whereas the *Soft* one undergoes the highest. Figure 5 reports the linear stress–strain chart for both synthetic and literature materials according to the constitutive model.

As can be seen in Figure 6, the strain map computed within the femoral cartilage after the simulation using the models with supports closely resembles that of anatomical models without supports, both under identical loading conditions. This indicates that the designed models with supports serve as effective physical simulators for studying the strain of synthetic cartilages.

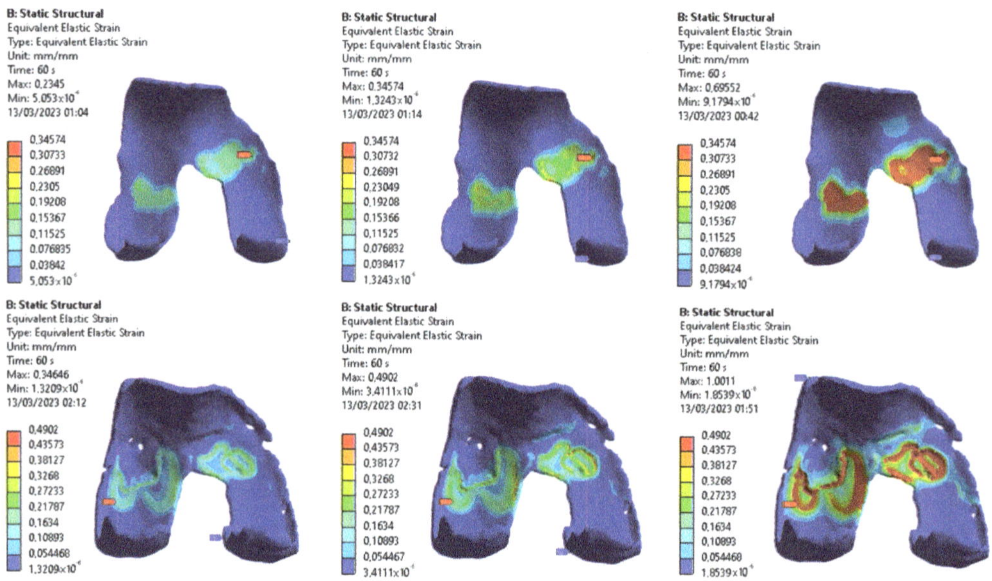

Figure 4. Strain map after 1150 N uni-axial load. Conditions in rows (*Control* and *Degenerative*, respectively) and materials in columns (*Medium*, *Literature*, and *Soft*, respectively). The color map was normalized on the values of the literature material.

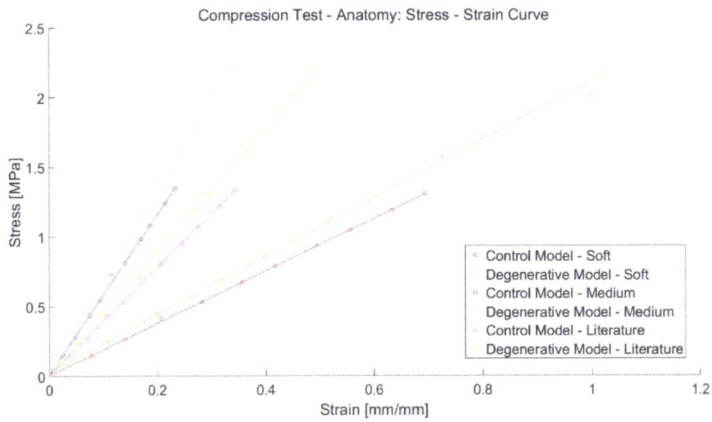

Figure 5. Stress-strain curves obtained after the simulation in ANSYS of a single-leg gait cycle in full-extended position utilizing the anatomical model without supports.

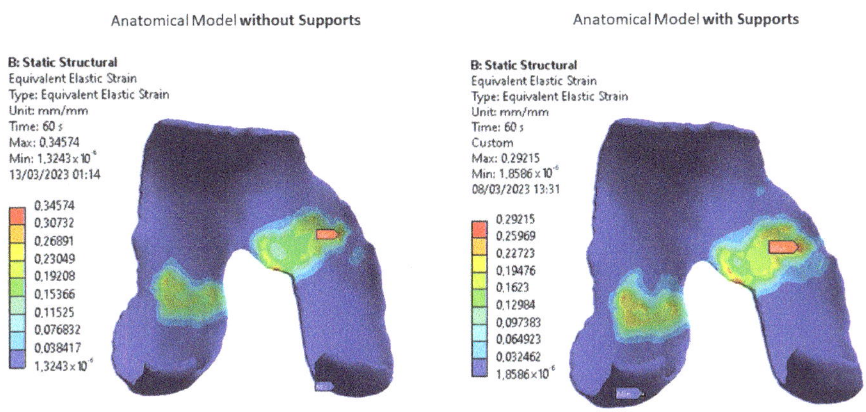

Figure 6. Strain distribution on healthy control cartilages following the computational simulation that differed between the two scenarios. On the left, anatomical models without supports were employed, whereas on the right, models designed with supports were utilized.

3.2. Mechanical Properties of Synthetic Polymers

The Young's modulus of *Soft* and *Medium* cartilage extracted from the mechanical test are, respectively, 2.43 MPa (R-square = 0.999 and RMSE = 0.01188) and 7.24 MPa (R-square = 0.9998 and RMSE = 0.1047).

Figure 7 reports the graphs related to the physical simulator setup that was the prototype during the study. The chart was reported as force–displacement since it was not feasible to measure the surface and the thickness of the contacting point. As can be seen, the Degenerative curves lie below the Control ones, despite the different materials. All tests were run until the 20% strain was reached and since the degenerative cartilage had a bigger volume, the final displacement was higher.

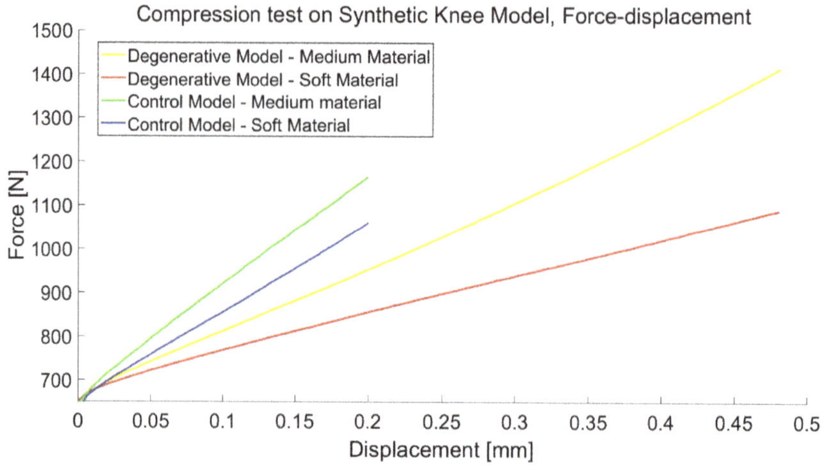

Figure 7. Force–displacement curves obtained from the compression test on the synthetic knee setup. It includes the results from the two patient-specific morphologies (Control and Degenerative).

Figure 8 reports the stress–strain curves of the customized specimens with fibers. Under uni-axial compression, the behavior of the 1% groups is similar, with a decreasing failure strain from the outside (*Superficial*) to the inside (*Deep*) of the cartilage. The increase in fibers volume content corresponded to an increase in the mechanical performance results: the stress is two times higher compared to the 1% group within the same strain range. Conversely, *Deep* 5% reported a nearly-linear behavior for small deformations, yielding a transaction around 0.1 MPa and a failure around 0.2 strain, making it the less flexible material.

The tensile behavior is linear for all six groups and also for the matrix itself (AGTM no fibers curve). This comparison was not possible under compression due to bending since the beginning of the test. Firstly, the presence of fibers increased the stress handled by the material but decreased the maximum elongation of the specimens. *Superficial* and *Middle* samples, both percentages, showed a similar trend, with *Superficial* having slightly higher stress. The presence of fibers aligned to the load direction led to an optimization of the structure and an increased response to a tensile load. The *Deep* family had the higher Young's modulus for both compositions. The elastic properties are reported in Table 3 for the compression tests and in Table 4 for the tensile tests.

Table 3. Compression Young modulus of fiber-reinforced samples, calculated on average for each group and expressed in MPa [N/mm^2].

Layer	1% Volume	5% Volume
Superficial	0.839	1.740
Middle	0.794	1.707
Deep	0.784	1.829

Table 4. Tensile Young modulus of fiber-reinforced samples, calculated on average for each group and expressed in MPa [N/mm^2].

Layer	1% Volume	5% Volume
Superficial	0.233	0.409
Middle	0.221	0.379
Deep	0.337	1.266

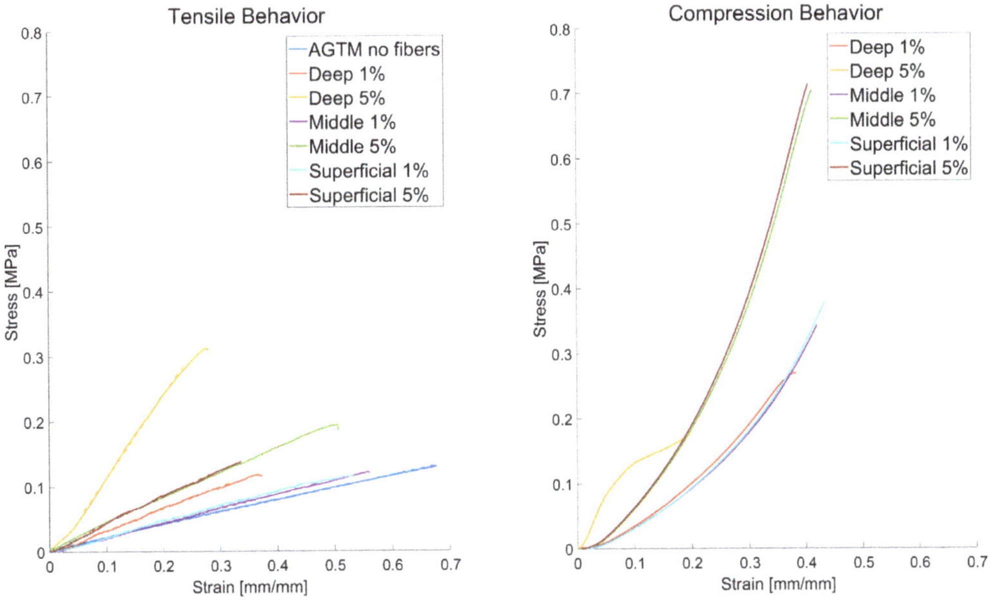

Figure 8. Stress-strain behavior for compressive and tensile tests of fiber-reinforced samples mimicking the three layers of native cartilage tissue. The tensile chart included a curve for the plain material (without fibers) to estimate the properties of the matrix itself.

4. Discussion

4.1. Radiological Data and FEA Simulation

Starting from patient-specific data, the segmentation workflow of medical images led to a reliable three-dimensional virtual model of the knee joint, with a focus on the main bones and cartilage tissue that compose it. This will support various clinical applications and allows for a better understanding of the knee biomechanics through a realistic 3D-printed model. Subject 1 (D), characterized by a greater volume, indicates potential fluid infiltration into the pathologic cartilage, leading to swelling and, thus, more intense strain [37]. In addition, the cartilage surface of Subject 2 appears to exhibit greater homogeneity compared to that of Subject 1 (D), accounting for the uneven distribution of holes and deformities in the latter.

Upon analyzing the Finite Element Analysis (FEA) results in Figure 5, it is evident that the stress–strain curves display a linear trend with remarkably similar slopes. The mechanical performance of biological cartilage tissue is intricately influenced by various factors, such as age, osteoarthritis (OA), and physical activity, making it challenging to establish a standardized mechanical profile [38]. Despite the distinct biochemical composition, the synthetic cartilage demonstrates promising overall elastic properties, thanks to the manual tunability achieved by adjusting the proportions of elastic or gel-like materials. This adaptability allows for the calibration of synthetic cartilage to match the desired reference values. The results position the printed materials as the endpoints on a spectrum, with *Soft* representing the lower limit and *Medium* representing the upper limit of the mechanical properties akin to knee joint cartilage (Figure 9). As stresses and deformations escalate, the impact of geometric variations becomes more pronounced, underscoring the role of osteoarthritis in modifying the mechanical response of cartilage [37]. These insightful comparisons underscore the potential of 3D printing and synthetic materials in effectively mimicking the elastic properties of natural cartilage tissue.

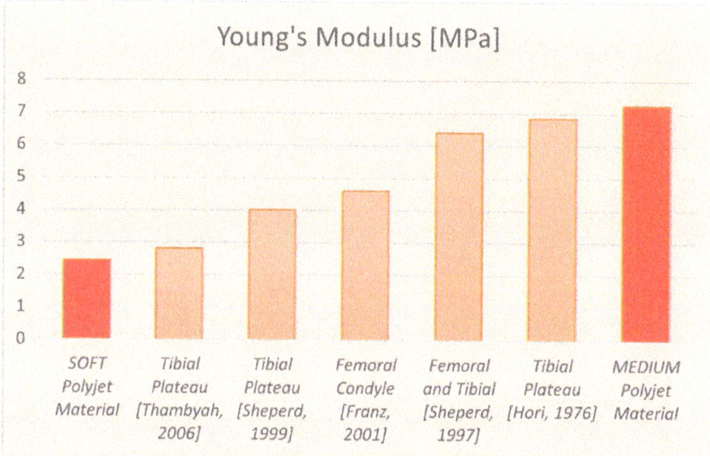

Figure 9. Young's modulus comparison of presented materials with different literature sources (from left to right: [39–43]). References come from healthy patients and different sites within the knee joint.

The FEM analyses carried out on the models with supports were used to verify that the created supports were actually able to simulate the correct distribution of loads in the knee. Despite surface modifications caused by the supports, the results of the stresses and strain distributions on the cartilages remained almost unchanged (Figure 6). Moreover, the materials' behavior in the stress–strain graphs remained consistent, affirming the minimal influence from the introduced supports on force transmission. These results validated the implementation of the mechanical setup with 3D-printed patient-specific phantoms shown in Figure 2.

Conventional testing techniques usually do not take into account tissue morphology and geometrical structure, in favor of concentrating only on the mechanical properties [44]. One significant simplification relates to the native cartilage's biomechanical behavior, which is bi-phasic poro-viscoelastic [45,46], but in this instance, it is assumed to be linear isotropic elastic. Nonetheless, several studies have found that this is suitable for instantaneous tissue deformation [26,28]. Moreover, other anatomical components of the knee joint, such as the tibial cartilages, menisci, and ligaments were not taken into account for the scope of this study. Even though these structures are essential to the health and function of the joint, their absence made it possible to gain a greater awareness of the behavior of pathological femoral cartilage and how it affects the health of the knee joint. This understanding permitted us to identify potential failure mechanisms associated with damage to the femoral cartilage.

4.2. *Synthetic Knee Joint Testing*

The curve behavior represented in Figure 7 is comparable to what is expected from the extrapolated parameters. The *Medium* material is stiffer, and thus a higher applied load is needed to reach 20% of the total deformation, compared to its soft counterpart; this applies for both control and degenerative morphologies, despite having different thicknesses and thus requiring more load in general to reach the deformation wanted. Overall, due to the curves' relative position, we can confirm that the presence of OA changes the response of cartilage to mechanical stimulation. In particular, the response is decreased for both materials, highlighting the role of geometry in the mechanics of the knee joint.

The results from FEA and synthetic testing align and show how morphology plays a crucial role in joint biomechanics, especially in the chain of degradation. The pathology causes the formation of brittle points; these degenerations leads to non-physiological load distributions that amplify the effect of the disease. The brittle points found in the FEA and in the synthetic tests are comparable and coincide with the failure areas in the synthetic setup. This means that a degenerative morphology is more suited to be implemented in this kind of parallel approach. And, in general, this leads to the conclusion that having a wider interacting surface on the virtual model is an important point in order to have comparable results to the mechanical test. For instance, OA cartilage tissue can be considered a perfect study case for this approach since it has been reported [20] to have an increased volume (and, thickness) due to swelling of the tissue, compared to the healthy physiological condition.

Furthermore, the stress on the tissue is unevenly distributed as can be observed in our analyses. The deformation is also increased in damaged areas since OA conditions bring wear and degradation. The strain distribution from the FEA (Figure 10) reveals the brittle regions of patient-specific cartilage during an extended-leg stride. As a first insight of this potential, we can qualitatively assess that the area experiencing higher deformations is much thinner and more convoluted in the degenerative knee, a consequence of the uneven load distribution between bones and cartilage.

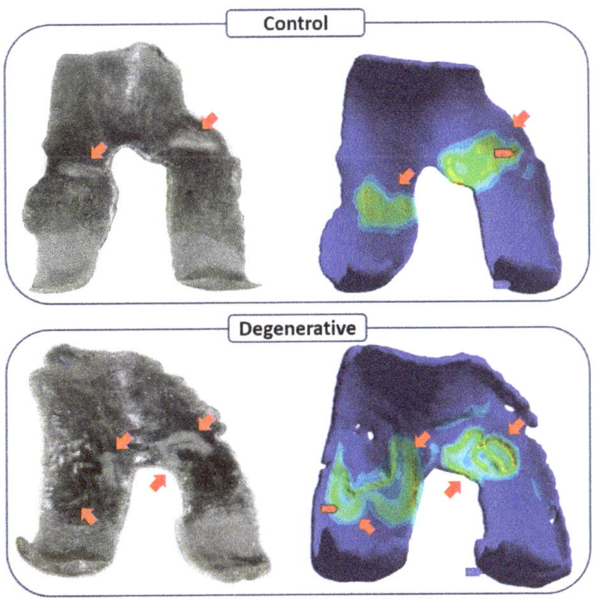

Figure 10. Qualitative assessment of brittle points on Control and Degenerative cartilages printed with *Medium* and *Soft* materials.

4.3. Fibers Contribute Evaluation

As shown in Figure 8, the mechanical behavior of fiber-reinforced samples is related to the spatial position of the samples. This corresponds with the literature findings demonstrating the highly anisotropic nature of cartilage, with Young's modulus exhibiting variations depending on the orientation of the fibers and direction of the applied load, thus highlighting a gradient [47,48].

The compressive resistance of cartilage tissue is guaranteed in the first place by the fiber's orientation in the *Superficial* sample. This is partially achieved in our samples: except for the 5% *Deep* samples, we can denote that this is true, with the *Superficial* samples showing the highest Young modulus. This is because fibers oriented tangentially to the applied force can prevent the cylindrical sample from barreling and bending, thus enhancing the resistance of the sample [47], even if the fibers' contribution seems to vary slightly among the different orientations when it comes to compressive properties.

From the tensile tests, the *Deep* samples show the highest Young modulus in both percentage configurations. In particular, the 5% *Deep* sample shows a greatly increased modulus compared to the others from the same group. This is an expected result, given the ability of fibers to enhance the tensile mechanical performance if oriented along the direction where the load is applied [48].

Again, by looking at the results of the compression and tensile tests, it is interesting to notice that the *Middle* samples show lower performance results for both fiber concentrations. It seems that a randomly oriented pattern of fibers yields a lower modulus compared to the parallel and tangential orientation. In the cartilage tissue, the middle layer acts as a transitional layer between the superficial and deep layers, ensuring an intermediate behavior between the two and helping to avoid abrupt compositional changes. Laasanen et al. [49] demonstrated that compressibility is lower in the middle zone, where the collagen II fibril network is less rigorously arranged. This aligns with our findings.

Overall, the elastic properties of our fiber-reinforced polymers have the same relationship observed in native cartilage layers, albeit with a lower absolute value.

5. Conclusions

In summary, while acknowledging the limitations of our 3D-printed cartilage, such as the challenges in replicating the exact biochemical composition and water content of natural cartilage, our primary focus was on assessing the potential of Digital Anatomy Printer (DAP) technology to emulate the mechanical behavior of knee cartilage. We explored the capability of DAP technology to modify overall mechanical responses through adjustments in polymer compounds and the manipulation of stiffness and elasticity. Our efforts also aimed at precisely tuning compressive and extension responses by varying fiber incorporation within the polymer compound, allowing the replication of specific load-bearing characteristics observed in natural knee cartilage.

To examine the knee mechanical behavior under osteoarthritis conditions, this study integrated 3D printing, advanced image-processing techniques, and computational simulations. The importance of considering the global morphology of cartilage tissue in designing knee osteoarthritis treatment plans was underscored in the first two parts of the study. Future studies and medical advancements may benefit from employing a synthetic knee joint model and computational simulation to comprehensively investigate the cartilage morphology and mechanical behavior in parallel, providing a more holistic understanding of osteoarthritis pathology.

In the final phase, the study assessed the feasibility of mimicking cartilage tissue's internal structure and components, utilizing fiber-reinforced material printed with PolyJet technology. The subsequent plan involves designing a multi-layer model that incorporates the native cartilage's zone micro-structure division, optimizing parameters through a more accurate mathematical model. This approach aims to enhance the understanding of Short Fiber-Reinforced Polymer Composites' mechanical performance, optimizing the fibers' volume fraction, and evaluating the impact of the fiber dimensions, thereby saving cost and time.

In conclusion, this work demonstrated the potential to replicate the patient-specific morphology and mechanical properties of biological tissues with 3D-printable polymers. The envisioned future application involves the use of 3D-printed osteoarthritis cartilage as a scaffold for bioinjection, offering promising prospects for advancements in medical treatment.

Author Contributions: Conceptualization, R.F. and P.G.; methodology, D.C., G.D. and R.F.; tests, R.F., G.B., D.C. and G.D.; formal analysis, D.C. and G.D.; investigation, D.C. and G.D.; writing—original draft preparation, D.C. and G.D.; writing—review and editing, G.B., F.K.C., R.F. and P.G.; supervision and funding acquisition, P.G. All authors have read and agreed to the published version of the manuscript.

Funding: This research was funded by EU Horizon Projects RESTORE grant ID: 814558.

Institutional Review Board Statement: Not applicable.

Informed Consent Statement: Not applicable.

Data Availability Statement: The data presented in this study are available on request from the corresponding author. The data are not publicly available due to privacy.

Conflicts of Interest: The authors declare no conflicts of interest.

References

1. Sophia Fox, A.J.; Bedi, A.; Rodeo, S.A. The basic science of articular cartilage: Structure, composition, and function. *Sport Health* **2009**, *1*, 461–468. [CrossRef]
2. Mansour, J.M. Biomechanics of cartilage. *Kinesiol. Mech. Pathomech. Hum. Mov.* **2003**, *2*, 66–79.
3. Zhang, Y.; Jordan, J.M. Epidemiology of osteoarthritis. *Clin. Geriatr. Med.* **2010**, *26*, 355–369. [CrossRef]
4. Long, H.; Liu, Q.; Yin, H.; Wang, K.; Diao, N.; Zhang, Y.; Lin, J.; Guo, A. Prevalence trends of site-specific osteoarthritis from 1990 to 2019: Findings from the Global Burden of Disease Study 2019. *Arthritis Rheumatol.* **2022**, *74*, 1172–1183. [CrossRef] [PubMed]
5. Buckwalter, J.A.; Martin, J.A.; Brown, T.D. Perspectives on chondrocyte mechanobiology and osteoarthritis. *Biorheology* **2006**, *43*, 603–609. [PubMed]
6. Roemer, F.W.; Demehri, S.; Omoumi, P.; Link, T.M.; Kijowski, R.; Saarakkala, S.; Crema, M.D.; Guermazi, A. State of the art: Imaging of osteoarthritis—Revisited 2020. *Radiology* **2020**, *296*, 5–21. [CrossRef] [PubMed]
7. Farr, J.; Cole, B.; Dhawan, A.; Kercher, J.; Sherman, S. Clinical cartilage restoration: Evolution and overview. *Clin. Orthop. Relat. Res.* **2011**, *469*, 2696–2705. [CrossRef] [PubMed]
8. Chu, C.R.; Szczodry, M.; Bruno, S. Animal models for cartilage regeneration and repair. *Tissue Eng. Part B Rev.* **2010**, *16*, 105–115. [CrossRef]
9. MacDonald, E.; Wicker, R. Multiprocess 3D printing for increasing component functionality. *Science* **2016**, *353*, aaf2093. [CrossRef]
10. Gargiulo, P.; Giménez, C.C.; Pirozzi, M.A.; Thórdarson, Á.; Ricciardi, C.; Forni, R.; Bjornsson, G.A. 20 years of 3D printing in surgical planning of distraction osteogenesis. In *Handbook of Surgical Planning and 3D Printing*; Elsevier: Amsterdam, The Netherlands, 2023; pp. 95–123.
11. Forni, R.; Agnarsdóttir, S.; Torfason, B.; Gargiulo, P. Heart surgery: Septal defect. In *Handbook of Surgical Planning and 3D Printing*; Elsevier: Amsterdam, The Netherlands, 2023; pp. 143–169.
12. Ventola, C.L. Medical applications for 3D printing: Current and projected uses. *Pharm. Ther.* **2014**, *39*, 704.
13. J850 Digital Anatomy Printer Website. Available online: https://www.stratasys.com/en/3d-printers/printer-catalog/polyjet/j850-digital-anatomy/ (accessed on 30 November 2023).
14. Severseike, L.; Lee, V.; Brandon, T.; Bakken, C.; Bhatia, V. Polyjet 3D printing of tissue-mimicking materials: How well can 3D printed synthetic myocardium replicate mechanical properties of organic myocardium? *BioRxiv* **2019**. [CrossRef]
15. Mandrycky, C.; Wang, Z.; Kim, K.; Kim, D.H. 3D bioprinting for engineering complex tissues. *Biotechnol. Adv.* **2016**, *34*, 422–434. [CrossRef] [PubMed]
16. Vijayavenkataraman, S.; Yan, W.C.; Lu, W.F.; Wang, C.H.; Fuh, J.Y.H. 3D bioprinting of tissues and organs for regenerative medicine. *Adv. Drug Deliv. Rev.* **2018**, *132*, 296–332. [CrossRef] [PubMed]
17. Wu, Y.; Kennedy, P.; Bonazza, N.; Yu, Y.; Dhawan, A.; Ozbolat, I. Three-dimensional bioprinting of articular cartilage: A systematic review. *Cartilage* **2021**, *12*, 76–92. [CrossRef] [PubMed]
18. Sun, Y.; Zhang, D.; Liu, Y.; Lueth, T.C. Fem-based mechanics modeling of bio-inspired compliant mechanisms for medical applications. *IEEE Trans. Med Robot. Bionics* **2020**, *2*, 364–373. [CrossRef]
19. Aubonnet, R.; Ramos, J.; Recenti, M.; Jacob, D.; Ciliberti, F.; Guerrini, L.; Gislason, M.K.; Sigurjonsson, O.; Tsirilaki, M.; Jónsson, H., Jr.; et al. Toward New Assessment of Knee Cartilage Degeneration. *Cartilage* **2023**, *14*, 351–374. [CrossRef] [PubMed]
20. Ciliberti, F.K.; Guerrini, L.; Gunnarsson, A.E.; Recenti, M.; Jacob, D.; Cangiano, V.; Tesfahunegn, Y.A.; Islind, A.S.; Tortorella, F.; Tsirilaki, M.; et al. CT-and MRI-based 3D reconstruction of knee joint to assess cartilage and bone. *Diagnostics* **2022**, *12*, 279. [CrossRef]
21. Maas, S.A.; Ellis, B.J.; Rawlins, D.S.; Weiss, J.A. Finite element simulation of articular contact mechanics with quadratic tetrahedral elements. *J. Biomech.* **2016**, *49*, 659–667. [CrossRef]
22. Minns, R.; Steven, F. The collagen fibril organization in human articular cartilage. *J. Anat.* **1977**, *123*, 437.
23. Muir, H.; Bullough, P.; Maroudas, A. The distribution of collagen in human articular cartilage with some of its physiological implications. *J. Bone Jt. Surg. Br. Vol.* **1970**, *52*, 554–563. [CrossRef]
24. Responte, D.J.; Natoli, R.M.; Athanasiou, K.A. Collagens of articular cartilage: Structure, function, and importance in tissue engineering. *Crit. Rev. Biomed. Eng.* **2007**, *35*, 363–411. [CrossRef] [PubMed]
25. Xia, Y.; Momot, K. *Biophysics and Biochemistry of Cartilage by NMR and MRI*; Royal Society of Chemistry: Cambridge, UK, 2016.
26. Trad, Z.; Barkaoui, A.; Chafra, M.; Tavares, J.M.R. *FEM Analysis of the Human Knee Joint: A Review*; Springer: Berlin/Heidelberg, Germany, 2018.
27. Armstrong, C.; Lai, W.; Mow, V. An analysis of the unconfined compression of articular cartilage. *J. Biomech. Eng.* **1984**, *106*, 165–173. [CrossRef] [PubMed]
28. Donzelli, P.S.; Spilker, R.L.; Ateshian, G.A.; Mow, V.C. Contact analysis of biphasic transversely isotropic cartilage layers and correlations with tissue failure. *J. Biomech.* **1999**, *32*, 1037–1047. [CrossRef] [PubMed]

29. Mak, A.; Lai, W.M.; Mow, V.C. Biphasic indentation of articular cartilage—I. Theoretical analysis. *J. Biomech.* **1987**, *20*, 703–714. [CrossRef] [PubMed]
30. Li, G.; Lopez, O.; Rubash, H. Variability of a three-dimensional finite element model constructed using magnetic resonance images of a knee for joint contact stress analysis. *J. Biomech. Eng.* **2001**, *123*, 341–346. [CrossRef] [PubMed]
31. Pena, E.; Calvo, B.; Martinez, M.; Doblare, M. A three-dimensional finite element analysis of the combined behavior of ligaments and menisci in the healthy human knee joint. *J. Biomech.* **2006**, *39*, 1686–1701. [CrossRef] [PubMed]
32. Blankevoort, L.; Huiskes, R. Validation of a three-dimensional model of the knee. *J. Biomech.* **1996**, *29*, 955–961. [CrossRef]
33. Bachtar, F.; Chen, X.; Hisada, T. Finite element contact analysis of the hip joint. *Med. Biol. Eng. Comput.* **2006**, *44*, 643–651. [CrossRef]
34. Haut Donahue, T.L.; Hull, M.; Rashid, M.M.; Jacobs, C.R. A finite element model of the human knee joint for the study of tibio-femoral contact. *J. Biomech. Eng.* **2002**, *124*, 273–280. [CrossRef]
35. Yang, Z. *Finite Element Analysis for Biomedical Engineering Applications*; CRC Press: Boca Raton, FL, USA, 2019.
36. Bendjaballah, M.; Shirazi-Adl, A.; Zukor, D. Finite element analysis of human knee joint in varus-valgus. *Clin. Biomech.* **1997**, *12*, 139–148. [CrossRef]
37. Peters, A.E.; Akhtar, R.; Comerford, E.J.; Bates, K.T. The effect of ageing and osteoarthritis on the mechanical properties of cartilage and bone in the human knee joint. *Sci. Rep.* **2018**, *8*, 5931. [CrossRef] [PubMed]
38. Peters, A.E.; Akhtar, R.; Comerford, E.J.; Bates, K.T. Tissue material properties and computational modelling of the human tibiofemoral joint: A critical review. *PeerJ* **2018**, *6*, e4298. [CrossRef] [PubMed]
39. Thambyah, A.; Nather, A.; Goh, J. Mechanical properties of articular cartilage covered by the meniscus. *Osteoarthr. Cartil.* **2006**, *14*, 580–588. [CrossRef] [PubMed]
40. Shepherd, D.; Seedhom, B. The 'instantaneous' compressive modulus of human articular cartilage in joints of the lower limb. *Rheumatology* **1999**, *38*, 124–132. [CrossRef] [PubMed]
41. Franz, T.; Hasler, E.; Hagg, R.; Weiler, C.; Jakob, R.; Mainil-Varlet, P. In situ compressive stiffness, biochemical composition, and structural integrity of articular cartilage of the human knee joint. *Osteoarthr. Cartil.* **2001**, *9*, 582–592. [CrossRef] [PubMed]
42. Shepherd, D.; Seedhom, B. A technique for measuring the compressive modulus of articular cartilage under physiological loading rates with preliminary results. *Proc. Inst. Mech. Eng. Part H J. Eng. Med.* **1997**, *211*, 155–165. [CrossRef] [PubMed]
43. Hori, R.Y.; Mockros, L. Indentation tests of human articular cartilage. *J. Biomech.* **1976**, *9*, 259–268. [CrossRef]
44. Patel, J.M.; Wise, B.C.; Bonnevie, E.D.; Mauck, R.L. A systematic review and guide to mechanical testing for articular cartilage tissue engineering. *Tissue Eng. Part C Methods* **2019**, *25*, 593–608. [CrossRef]
45. Setton, L.A.; Zhu, W.; Mow, V.C. The biphasic poroviscoelastic behavior of articular cartilage: Role of the surface zone in governing the compressive behavior. *J. Biomech.* **1993**, *26*, 581–592. [CrossRef]
46. Eschweiler, J.; Horn, N.; Rath, B.; Betsch, M.; Baroncini, A.; Tingart, M.; Migliorini, F. The biomechanics of cartilage—An overview. *Life* **2021**, *11*, 302. [CrossRef]
47. Petitjean, N.; Canadas, P.; Royer, P.; Noël, D.; Le Floc'h, S. Cartilage biomechanics: From the basic facts to the challenges of tissue engineering. *J. Biomed. Mater. Res. Part A* **2023**, 111, 1067–1089. [CrossRef] [PubMed]
48. Antons, J.; Marascio, M.G.M.; Nohava, J.; Martin, R.; Applegate, L.; Bourban, P.; Pioletti, D. Zone-dependent mechanical properties of human articular cartilage obtained by indentation measurements. *J. Mater. Sci. Mater. Med.* **2018**, *29*, 1–8. [CrossRef] [PubMed]
49. Laasanen, M.; Töyräs, J.; Korhonen, R.; Rieppo, J.; Saarakkala, S.; Nieminen, M.; Hirvonen, J.; Jurvelin, J. Biomechanical properties of knee articular cartilage. *Biorheology* **2003**, *40*, 133–140. [PubMed]

applied sciences

MDPI

Article

Assessment of Surrogate Models for Research on Resistance and Deformation of Repairs of the Human Meniscal Roots: Porcine or Older Human Models?

Alejandro Peña-Trabalon, Ana Perez-Blanca, Salvador Moreno-Vegas, M. Belen Estebanez-Campos and Maria Prado-Novoa *

Laboratory of Clinical Biomechanics of Andalusia (BIOCLINA), 29071 Málaga, Spain;
alejandropt98@uma.es (A.P.-T.); anaperez@uma.es (A.P.-B.); belen@uma.es (M.B.E.-C.)
* Correspondence: maria.prado@uma.es

Featured Application: This study provides data and rationales to aid researchers in the selection of surrogate models for in vitro assessments of surgically repaired meniscal roots. The results are potentially applicable in experimental in vitro investigations aimed at evaluating the performance of surgical repair both in existing approaches and emerging techniques. It could also be of interest for studies seeking to adjust material models of the meniscal tissue around the suture area for incorporation into computational models.

Abstract: Meniscal root repair is not routinely recommended for patients over 75 years old, yet surrogate age-unrestricted human or porcine models are used for its evaluation. This study assesses the suitability of older human or porcine meniscus models for in vitro testing of the sutured meniscal horn. Three groups of menisci underwent a load-to-failure test with continuous monitoring of the traction force and deformation around the suture: human < 75 years, human ≥ 75 years, and porcine. Both surrogate models were compared to the younger group. The porcine group exhibited a 172.1%-higher traction force before tearing ($p < 0.001$) and a 174.1%-higher ultimate force ($p < 0.001$), without there being differences between the human groups. At tissue level, the older group had a 28.7%-lower cut-out stress ($p = 0.012$) and the porcine group had a 57.2%-higher stress ($p < 0.001$). Regarding elasticity at the sutured area, a 48.1%-greater deformation rate was observed in the older group ($p < 0.001$), without difference for the porcine group. In conclusion, neither the porcine nor the older human model demonstrated a clear advantage as a surrogate model for young human sutured meniscal horns. The older human meniscus is preferable for resistance at the specimen level, while the porcine model better represents deformation in the sutured zone.

Keywords: meniscal repair; surrogate models; human meniscal tissue; meniscal root detachment; suture; resistance; deformation

1. Introduction

The menisci of the knee are fixed by ligamentous attachments that retain them in the intra-articular zone while allowing the necessary mobility (up to 10 mm [1]) to be placed between the tibia and femur at the contact area, which shifts during flexion–extension. The meniscal roots are the only ligamentous attachments with direct insertion into the bone, playing a crucial role in meniscal function. Complete meniscal root avulsion has been shown to cause significant changes in the contact biomechanics of the knee, increasing pressure and decreasing the contact area on the cartilage of the affected compartment, changes that may be similar to those observed for total meniscectomy [2–4]. Biomechanical studies have also highlighted the role of meniscal roots as secondary stabilizers of knee kinematics, contributing to the restriction of anteroposterior translation and internal–external rotation [5–7]. From a clinical point of view, a meniscal root detachment can lead to the rapid development of arthritis [8] or osteonecrosis [9] as a consequence of biomechanical alterations. The surgical treatment of meniscal root avulsions has currently evolved from partial meniscectomy to meniscal root reinsertion [10]. The surgical repair can be performed using two techniques: transtibial [11,12] or in situ fixation [4,13]. In both approaches, sutures are passed through the horn of the injured meniscus to hold it in place.

Some research studies have subjected sutured meniscal horn specimens to in vitro tests to evaluate the performance of surgical suture materials [14–17], the effectiveness of suture techniques [12,18–21], the validity of suturing or fixation devices [21–24], and the suitability of surgical approaches [18,25–27]. Both the resistance of the meniscal horn to suture traction without tear initiation and deformation at the suture area are assessed in these studies. Knowledge of the resistance to traction of the repaired meniscus is important in preventing the failure of surgical intervention, while monitoring post-repair displacements generated by suture traction is needed to ensure that they remain within clinically acceptable limits [28].

Porcine models are often chosen in experimental constructs [12,15–19,21,23,26–29] because of their reduced variability and their similarities with human menisci in terms of anatomical structure, vascularity, volume, and weight, although the width of porcine menisci is greater [30]. In addition, the kinematics of the porcine knee reasonably approximates that of the human knee. Consequently, the porcine model is a practical and economically feasible choice that finds extensive use in the biomechanical testing of meniscal repair and replacement techniques and allows for the reasonable translation of the results to clinical applications [31–34]. However, it is not proven that the entire set of biomechanical properties of the porcine meniscus matches those of the human meniscus [33–35], raising concerns about the use of this model as a human surrogate and underscoring the need to evaluate its applicability in relation to the specific parameters intended to be studied.

Other in vitro studies use human menisci [4,36–39], despite their greater heterogeneity among donors. And perhaps more importantly, although meniscal root repair is indicated for non-elderly patients, with the age cut-off being increasingly high in new paradigms [40–42], donors are not limited by age in many in vitro biomechanical studies. Therefore, the studies sometimes involve tissue from donors of a very advanced age.

Another interesting use of experimental models of the meniscus is in computational modeling, where material models of meniscal tissue are adjusted from properties obtained from in vitro tests of tissue samples from various origins, including animals and older humans. In this context, the analysis and optimization of different suturing techniques for meniscal root repair could greatly benefit from a tissue model that incorporates the effect induced by thread traction around the suture hole. In order to have an adequate model, it is necessary to perform a large number of experiments, with tissue availability being a major issue.

It is currently unknown as to whether the use of porcine or aged cadaveric material is appropriate for studying the behavior of repaired meniscal roots in adults. To our knowledge, their suitability as substitutes for younger human tissue has not been compared. The objective of this study is to assess the suitability of these two models for in vitro testing of meniscal root repair in terms of suture pull-out resistance and suture traction-induced displacements around the suture site. Our working hypotheses in studies on the mechanical behavior of the sutured meniscal horns in adults are the following: (1) meniscal tissue from older donors ($\geq$75 years old) does not adequately represent the biomechanical behavior of the younger meniscus; (2) the porcine meniscus is a good surrogate for the human meniscus at an age routinely eligible for repairing.

2. Materials and Methods

To test the working hypotheses, three groups of isolated sutured menisci were considered: human <75 years, human $\geq$75 years, and porcine. The human <75 years group acted as the control group, while the human $\geq$75 years and porcine groups represented the experimental models under analysis. Isolated meniscus models were tested to focus the study on the behavior of the tissue–suture interface.

After approval by the Ethical Committee of Experimentation of the University of Malaga, to extract the menisci, 38 cryopreserved human knees from the mid-femur to the mid-tibia with no previous history of knee osteoarthritis, knee fractures, or surgery on the meniscus or tibial plateau, provided by a specialized company, and 22 stifle joints of adult pigs donated by a Spanish local slaughterhouse, also from the mid-femur to the mid-tibia, were used, complying with all legal and ethical requirements.

2.1. Specimen Preparation

Adult human knee joints were classified in two age groups, with an age limit of 75 years old. They were individually stored at a temperature of −20 °C in sealed plastic bags. The age limit of 75 years old was selected, taking into account recent recommendations on age extension for meniscal repair surgery.

The day prior to testing, the knee was left to thaw at room temperature, wrapped in dampened gauze. Once thawed, the knee was dissected, and the menisci were extracted. They were visually inspected for any potential damage or pathology, specifically at the roots and horns. The eligibility criteria required a minimum macroscopic quality grade of 3, according to the scale of Pauli et al. [43]. Upon meeting this criterion, the menisci were randomly assigned to undergo either anterior or posterior surgery. Finally, each meniscus was wrapped in a dampened gauze and individually packed in a sealed plastic bag to keep it hydrated in the cooler until the time of testing. Sixty-six human menisci, divided into two groups, were included in the study.

Porcine menisci were harvested, preserved, and prepared following the same procedure described for human specimens. The samples were obtained from skeletally mature, freshly slaughtered 6-month-old pigs weighing approximately 100 kg. The required inclusion criteria were the absence of macroscopic degenerative changes and traumatic damages.

At the time of testing, the meniscus was taken from the freezer and removed from the plastic bag. Then, a surgical suture of the horn was simulated by a specialized surgeon inserting a #2 non-absorbable, high-resistance, 100% UHMWPE, braided fiber thread (Force Fiber™ #2, Stryker Iberia, Madrid, Spain) using the attached $\frac{1}{2}$ circle tapered needle. A single simple suture was chosen for the surgical simulation. The puncture was made at 5 mm from the internal meniscal edge and from the end of its root to be consistent with the zone where the surgical hole is made during the surgical procedure [29,44,45]. Although it is known that a single simple stitch is neither the most resistant nor the most commonly applied method for repairing a complete meniscal horn tear, it was selected because the use of multiple threads introduces uncertainty in load distribution, which hinders accurate load determination at the meniscus–suture interface. The use of a single suture creates the simplest tissue–suture interface, making it easier to quantify the resistant area of the tissue.

After suturing, the meniscus thickness was measured with a manual caliper at the point where the needle had been inserted. Then, the suture–meniscus set was wrapped again in a dampened gauze to keep it hydrated.

2.2. Biomechanical Testing

A single-axis testing machine (Figure 1), specially designed for biomechanical tests [46], was used to subject the sutured meniscus to a load-to-failure test. Three orthogonal axes are defined in the testing machine, with the z-axis coinciding with the servomotor screw that sets the direction of traction (Figure 1). The suture–meniscus set was placed in the testing machine in two phases. First, the meniscus was fixed with a clamp at approximately 8 mm from the suture point, using sandpaper surrounding the clamping area to increase friction. The meniscus was carefully oriented with the cranial surface facing the outer side of the testing machine, the longitudinal fibers of the horn were aligned to the loading direction, and the suture hole was centered with the actuator head in the loading direction. Second, the two free ends of the suture were also wrapped in sandpaper and attached to a mechanical clamp on the machine head, with 55 mm between the puncture point and the limit of the clamp when the suture is pulled manually, with just enough traction to keep it vertical. Fifty-five millimeters were selected as representative of the expected suture length in a transtibial meniscal horn repair [14,47]. An electronic inclinometer (Bubble Level 3D, v. 2.2.4, Maleirbag, 2022, Puebla, Mexico, installed on the smartphone M2002F4LG, Xiaomi, Beijing, China) was used to orient the meniscus so that its cranial and caudal surfaces were within the plane defined by the *z*-axis (traction direction) and the *y*-axis (transverse direction) of the testing machine as close as possible. Once the meniscus–suture set had been placed on the testing machine, two ink points were marked with a surgical pen on the cranial side. Aided by the inclinometer to identify the direction of traction on the meniscal surface, the points were located along a line in that direction passing through the suture hole: Point1 on the suture limb coincident with the meniscus–suture interface; and Point2 on the opposite side of the hole in its immediate surroundings (Figure 2). These marks were used for videogrammetric analysis of the displacements around the hole in the direction of traction.

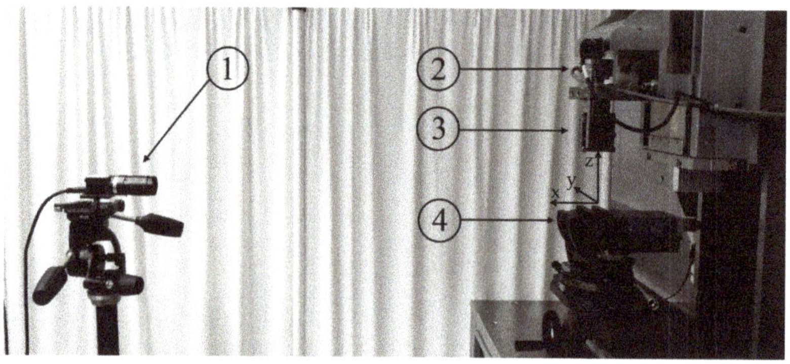

Figure 1. Uniaxial testing machine and the videogrammetric system: (1) digital camera; (2) load cell; (3) clamp on the head of the testing machine to fix the suture; (4) clamp on the base of the testing machine to fix the meniscus.

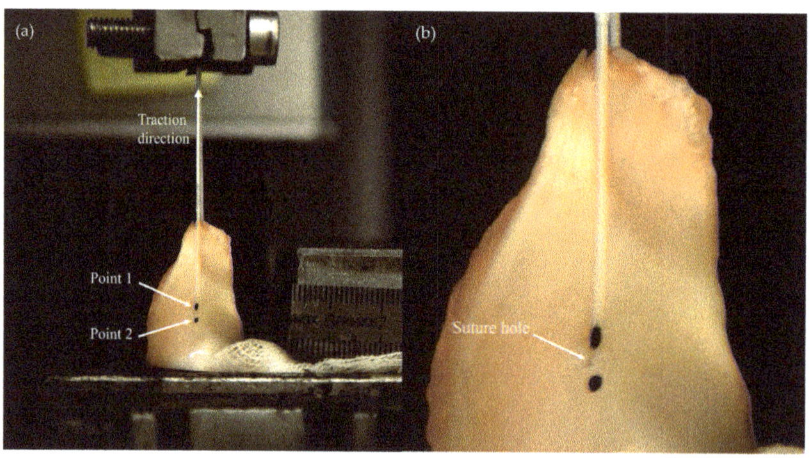

Figure 2. (**a**) Meniscus specimen on the testing machine. The marks for videogrammetric tracking can be observed. (**b**) Detail of the suture hole area.

The servo controller of the testing machine (SGDH-15AE-S-OY, Yaskawa Electric, Fukuoka, Japan) recorded the actuator displacement in the pull direction, and a 2000 N load cell of accuracy class 0.1 (U2B, HBM, Darmstadt, Germany) registered the applied traction force. Both signals were sampled at 1000 Hz. For videogrammetric analysis, a digital camera (VCXU-124C, Baumer, Frauenfeld, Switzerland) (Figure 1), synchronized with the sensors of the testing machine, was placed facing the cranial surface of the meniscus. For this purpose, the camera was mounted on a tripod (808RC4, Manfrotto, Cassola, Italy) featuring three rotating orthogonal mechanisms (Figure 1), and, using the electronic inclinometer, it was positioned to align its image plane with the plane containing the z and y axes of the testing machine, i.e., approximately parallel to the mid transverse plane of the meniscus. The camera was equipped with a 2.8/75 mm lens (C7528-M, Pentax, Tokyo, Japan). Images were acquired with a 250 ms sampling period.

The sutured meniscus was initially pulled to 1 N at 0.05 mm/s, held at this load level for 5 s to allow the specimen to stabilize, and then subjected to a displacement-controlled load-to-failure test at 0.1 mm/s.

2.3. Data Analysis

To assess the resistance at the specimen level, the force that initiated this meniscal tear, i.e., the meniscal cut-out force, F_C, was registered for each specimen. The initiation of meniscal tissue cut-out was identified as the point of change in the slope of the curve representing the displacement of Point1 with respect to Point2, D, as a function of the traction force [46]. The tearing initiation was also verified by checking the video images to validate the accuracy of the detection. The maximum force borne by the specimen in the load-to-failure test, i.e., the ultimate force, F_u, was also registered.

At the tissue level, the tissue cut-out stress of the meniscal horn, S_c, was calculated as the engineering stress at the suture–meniscal interface at the tearing initiation as follows:

$$S_c = \frac{F_C}{\phi \cdot h}, \quad (1)$$

where ϕ is the nominal diameter of the suture, and h is the meniscal thickness at the hole, i.e., the denominator is the projected area at the meniscus–suture interface. Equation (1) may underestimate the contact surface at the beginning of the test since the suture passage may not be perfectly orthogonal to the meniscal surfaces despite being subjected to a tensile force of 1 N. However, as the test progresses, the thread will rapidly orient itself to minimize the contact area.

To evaluate tissue deformation around the suture site, an equivalent stiffness modulus on the traction direction of the suture, m, was calculated as the slope of the linear approximation of the stress–strain curve, $\sigma - \varepsilon$, in the range of $\varepsilon = [0,\ 0.3]$. The engineering stress in the traction direction at the meniscus–suture interface was calculated for the test points prior to the initiation of tearing with an expression similar to the one used in Equation (2), i.e.,

$$\sigma = \frac{F}{\phi \cdot h}, \quad (2)$$

where F is the traction force recorded by the load cell during the test. And the strain was computed as follows:

$$\varepsilon = \frac{D}{D_0} - 1, \quad (3)$$

where D_0 is the distance between Point1 and Point2 computed in the first frame of the load-to-failure test, and D is the distance between them computed in subsequent frames.

2.4. Statistical Analyses

Sample size was determined using G*Power 3.1.9.7 software [48] for a minimum detectable difference of 15 N between the older human group and the younger human group as the control group, using a t-test at $\alpha = 0.05$ with a power of 0.80 and an allocation ratio of $N/N_c = 0.5$ and a standard deviation of 21 N estimated from a preceding work [49].

The resulting minimum control group size was $N_c = 41$. The difference of the 15 N selected represents 25% of the F_c of the 60 N estimated for the younger human group from the data of the aforementioned previous work. This difference was considered sufficiently large so the adequacy of the experimental models analyzed, i.e., the human ≥75 years and porcine groups, could not be considered representative of the control group, i.e., human <75 years. Finally, conservative group sizes of $N_c = 44$ for the control group and $N = 22$ for the test groups were chosen to allow for the exclusion of specimens and samples that did not meet the criteria for parametric testing.

Descriptive statistics and comparative tests were conducted using IBM® SPSS® Statistics, v.23 (IBM, Chicago, IL, USA). To assess differences in the mechanical properties at the suture site between the younger human group and each of the groups representing a different experimental model, non-parametric Kruskal–Wallis omnibus tests for independent samples were carried out. When differences were detected, post hoc analysis consisted of pre-planned comparisons of each model with the control group using Dunn–Bonferroni tests with correction for two comparisons. These nonparametric tests were chosen because all samples failed either the normality test or the homogeneity of variances test. p values ≤ 0.05 were regarded as significant.

3. Results

The three study groups included in the study showed the following characteristics:

- Human <75 years group: 45 younger-than-75-years-old menisci (mean age 51.00 years, SD 13.56, range 28–67 years) obtained from 24 cadaveric knees of 24 donors after discarding 3 menisci due to not meeting the inclusion criteria. The group consisted of 27 menisci from men and 18 from women, 23 medial and 22 lateral menisci, 22 for anterior and 23 for posterior surgery.
- Human ≥75 years group: 21 75-years-old-or-older specimens (mean age 84.86 years, SD 3.82, range 82–95 years) obtained from 14 cadaveric knees of 14 donors after discarding 7 menisci due to not meeting the inclusion criterion. The group consisted of 9 menisci from men and 12 from women, 11 medial and 10 lateral menisci, 9 for anterior and 12 for posterior surgery.
- Porcine group: 21 menisci from 22 6-month-old pigs. All of them met the inclusion criteria. The group consisted of 11 medial and 11 lateral menisci, 10 for anterior and 12 for posterior surgery.

3.1. Meniscal Horn Thickness at the Suture Point, *h*

At the suture site, older human menisci were thicker than the younger menisci ($p = 0.012$), with a 24.7%-greater mean value. The difference in thickness was even more pronounced in the porcine group, which was 83.8% thicker ($p < 0.001$) (Figure 3 and Table 1).

Table 1. Meniscus thickness of the meniscal horn at the sutured area.

	Human <75 Years	Human ≥75 Years	Porcine
Mean	2.71 mm	3.38 mm ‡	4.98 mm ***
SD	0.60 mm	0.85 mm	1.13 mm

Significant difference compared to human <75 years: ‡ $p = 0.02$; *** $p < 0.001$.

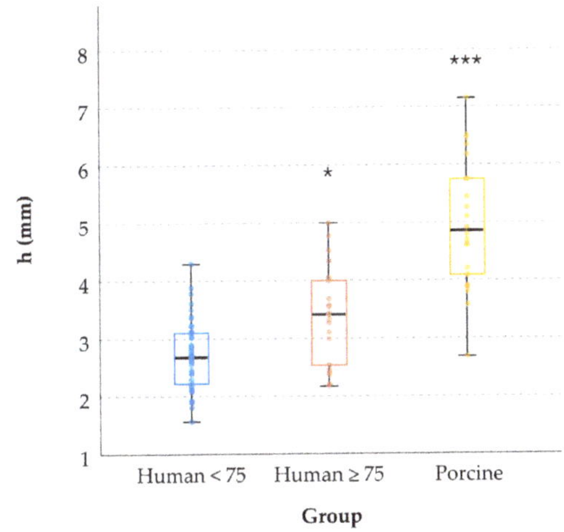

Figure 3. Box plot showing the meniscal horn thickness at the sutured area, h, for the test groups. Each box represents the interquartile range (IQR) of the respective group, i.e., the 25–75% percentile, with the central horizontal line indicating the median value. The upper and lower limits of the vertical line correspond to the maximum and minimum values excluding outliers, i.e., the points outside 1.5 times the IQR. Dots represent all data points. Significant difference compared to human <75 years are indicated as * $p < 0.05$; *** $p < 0.001$.

3.2. Meniscal Resistance to Cut-Out Initiation, Fc

Compared to the younger human group, the specimen resistance to cut-out initiation by suture traction, F_C, was significantly higher in the porcine group ($p < 0.001$), with a 172.1% increase in the mean value (Table 2a and Figure 4a). However, no significant difference was found between the younger and older human groups.

Table 2. Mechanical properties of the meniscal horns at the suture insertion point area: (a) specimen cut-out force; (b) specimen ultimate force; (c) tissue cut-out resistance; (d) tissue equivalent stiffness modulus.

			Human <75 Years	Human ≥75 Years	Porcine
(a)	F_c (N)	Mean	60.1	54.5	168.9 ***
		SD	19.9	16.9	50.4
(b)	F_u (N)	Mean	75.1	70.5	205.8 ***
		SD	29.6	21.9	74.3
(c)	S_c (MPa)	Mean	47.4	33.8 *	74.5 ***
		SD	14.1	13.1	27.5
(d)	m (MPa)	Mean	110.7	57.5 ***	111.3
		SD	55.6	32.0	52.1

Significant difference compared to human <75 years: * $p < 0.05$, *** $p < 0.001$.

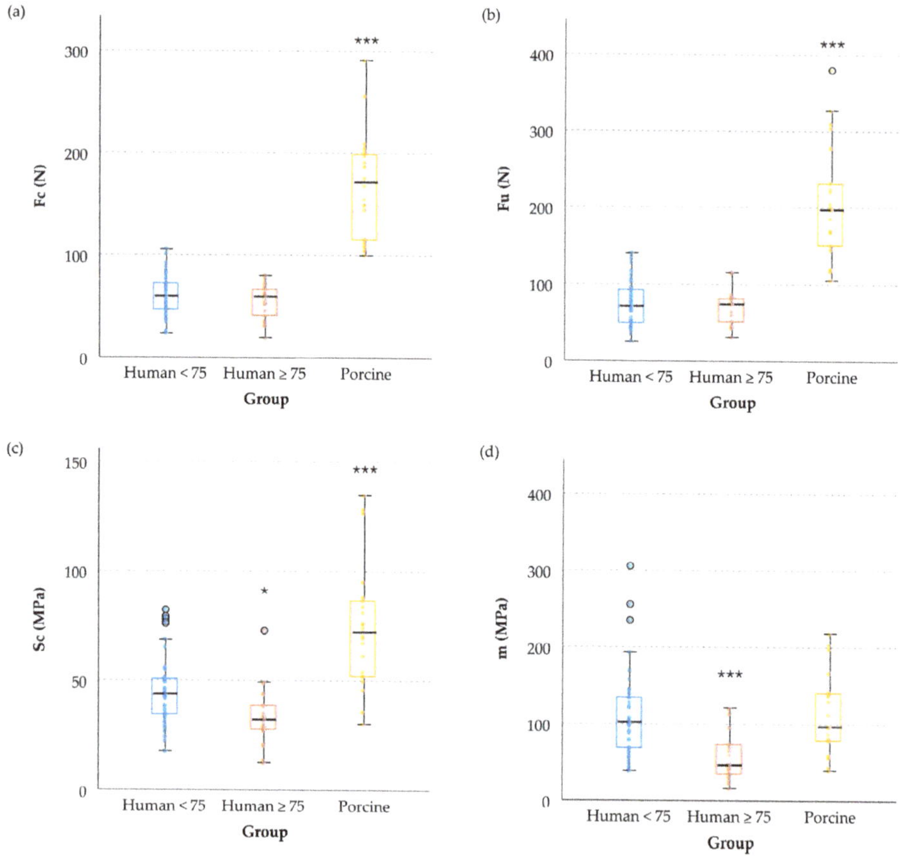

Figure 4. Box plots of the test groups showing (**a**) specimen cut-out force, F_C; (**b**) specimen ultimate force, F_u; (**c**) tissue cut-out resistance, S_C; (**d**) tissue equivalent stiffness modulus, m. For each group, the box represents the interquartile range (IQR) of the respective group, i.e., 25–75% percentile, with the central horizontal line indicating the median value. The upper and lower limits of the vertical line correspond to the maximum and minimum values excluding outliers, i.e., the points outside 1.5 times the IQR. Small circles correspond to outliers, if found. Dots represent all data points. Significant difference compared to human <75 years are indicated as * $p < 0.05$; *** $p < 0.001$.

Analysis of the video recordings confirmed that tearing was initiated slightly before the first local maximum of the force–deformation curve, consistent with previous reports [46].

3.3. Meniscal Ultimate Force, Fu

The results of the comparison between the testing groups and the control for the ultimate meniscal force were similar to those for F_c (Table 2b and Figure 4b): the older human group was quite similar to the younger human group (the mean values differed by only by 6.1%), while the porcine model was 174.1% larger ($p < 0.001$).

Comparing the mean values of the ultimate force and the resistance to cut-out initiation for each group, F_u was 24.9% higher than F_c in the younger human group, 29.3% in the older human group, and 21.9% higher in the porcine model.

3.4. Tissue Cut-Out Resistance, Sc

When comparing the human meniscus groups, the sutured tissue in the older group was less resistant, with a mean S_C 28.7% lower than in the younger group ($p = 0.012$). On the other hand, porcine tissue was more resistant than younger human tissue ($p < 0.001$), requiring a 57.2%-higher mean stress value to reach the cut-in point (Table 2c and Figure 4c).

3.5. Tissue Equivalent Stiffness Modulus, m

figfig:applsci-2773800-f005a illustrates typical load–displacement curves for each of the sample types included in the study. Figure 5b displays the corresponding stress–strain curves for these representative samples, along with their linear adjustments in the strain interval [0, 0.3], and indicates the adjusted R-squared values of the fittings. Similar to the curves in Figure 5b, all meniscal horns showed highly linear behavior in the strain range [0, 0.3]. Specifically, the adjusted R-squared values for the linear fitting were within the range [0.91, 0.99] for human <75 and human ≥75 models, and in the range [0.92, 0.99] for the porcine model. As justified in the Materials and Methods section, the equivalent stiffness modulus, m, was computed as the slope of the aforementioned linear adjustments, resulting in the mean values in Table 2d.

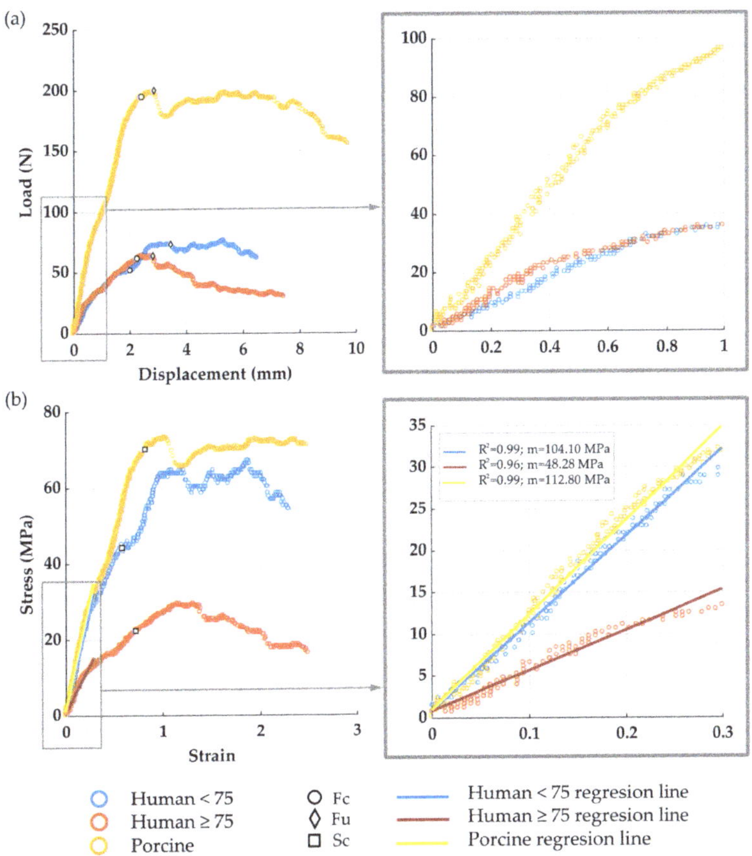

Figure 5. Typical curves for each of the sample types included in the study: (**a**) load–displacement curves with magnification of the displacement interval [0, 1]; (**b**) corresponding stress–strain curves with magnification of the strain interval [0, 0.3], along with their linear fitting within this interval, indicating the adjusted R-squared values.

Meniscal tissue at the sutured area was more elastic in the older group than in the younger group ($p < 0.001$ for both comparisons), with m being 48.1% greater (Table 2d and Figure 4d). In porcine specimens, the elasticity was very similar to younger human menisci, with almost identical mean m values (difference less than 0.6%).

4. Discussion

The main result of our study is that neither the porcine model nor the human model with cadaveric tissue from donors of advanced age (over 75 years old) can provide accurate representations of all the mechanical properties of adult human specimens under 75 years old. However, while neither of the two models offers a clear advantage for a general use in biomechanical in vitro tests, each could be adequate for examining specific characteristics of the sutured meniscal horn.

Regarding the resistance to suture-induced traction at the specimen level, no difference was observed between the older and younger human meniscus groups, which partially contradicts our first hypothesis that meniscal tissue from older donors (≥75 years) does not adequately represent that of the younger menisci. On the other hand, the porcine model was significantly more resistant than the younger human model at the specimen level, with differences of 172.1% in F_c and 174.1% in F_u, as opposed to our second hypothesis. Upon examining resistance at the tissue level in terms of tissue cut-out stress at the meniscus–suture interface, S_c, both models exhibited significant differences from the younger human group. Specifically, porcine specimens exhibited tissue with a 57.2%-greater resistance to suture-induced tearing, whereas specimens from donors over 75 years old showed less-resistant tissue, with a 28.7% reduction compared to the human meniscus from younger donors. Regarding deformation, the findings from the comparison between the models opposed those observed in specimen resistance. In particular, the equivalent stiffness modulus, m, of the porcine model did not differ from that of the younger human models, while the human ≥75 years model showed a 48.1% increase in this elastic parameter compared to the human meniscus <75 years.

Consequently, the adequacy of these two models depends on the interest of the study, a result that partially contradicts both of our initial hypotheses that the meniscal tissue from older donors (≥75 years) does not adequately represent that of the younger menisci and that the porcine meniscus is a good surrogate for the human meniscus at an age routinely eligible for repairing. In research on mechanical properties at the specimen level, to analyze the resistance of sutured meniscal roots, the older human meniscus model resulted to be a preferable surrogate model to the porcine model; conversely, if the focus is on issues related to deformation before cut-out initiation in the sutured root zone, the porcine model offers a more accurate representation. For studies at the tissue level, as experimental tests to determine material properties around the sutured root area, neither model resulted to be a reliable surrogate in terms of resistance; but in tests where only tissue deformation prior to cut-out initiation is of interest, opting for the porcine model may be preferable. However, caution is advised in the latter case, as the tissue stresses would be inadequately characterized.

The differences found in the suitability of the surrogate models when used in studies evaluating meniscal root resistance at the specimen level or tissue level may be partially attributed to differences in thicknesses observed among the groups. The porcine model offers a surrogate for the human meniscus <75 years, which has a more resistant tissue, with a 57.2% higher S_c. This difference is further accentuated at the specimen level due to the thicker root of the porcine meniscus, resulting in a 172.1%-higher F_c and a 174.1%-higher F_u. Conversely, when employing an older human meniscus model as a surrogate, although the tissue is less resistant, with a 28.7%-lower S_c, this effect is offset at the specimen level. Likely, the reason is that the meniscal root of the older group was thicker than that of the younger group, resulting in the vanishing of differences at the specimen level.

Large animal models are valuable tools in biomechanical engineering, finding applications in both preliminary in vitro testing and in vivo testing. These models, including dogs, rabbits, cows, sheep, and pigs, have been extensively utilized in research on the natural meniscus, the study of its kinematic behavior, the assessment of the biomechanics of knee joint contact, the validation of engineered tissues proposed for total or partial replacement of the meniscus, and the evaluation of the efficacy of surgical meniscal repairs. The canine model, historically employed in studies, has experienced a decline in usage, potentially due to increasing pressure from animal advocacy groups. Moreover, from a technical perspective, the majority of surgical procedures on the canine meniscus require arthrotomy, often accompanied by medial collateral ligament rupture, deviating from arthroscopic procedures in humans. The rabbit model exhibits significant differences in vascularity, collagen orientation, and glycosaminoglycan content of the meniscus compared to human menisci. Additionally, the kinematics of the rabbit's femorotibial joint significantly differ from those of the human knee [31,34,50]. The bovine model, significantly larger than the human meniscus, has limited application in biomechanical studies due to the size discrepancy [34,50]. The ovine meniscus, while anatomically similar to the human meniscus, differs in kinematics and contact force distribution, making it unsuitable for non-isolated studies of the repaired meniscus [30,31,34]. Regarding the porcine model, it closely resembles the human model in terms of anatomical structure, vascularity, volume, and weight [30,31,34], although the width of the meniscus is greater, as demonstrated in the present study. The kinematics of the porcine knee reasonably approximate those of the human knee. As a result of the aforementioned arguments, the porcine model is the predominant surrogate for in vitro tests of the young human meniscus [12,15–19,21,23,26–29], widely applied in biomechanical testing for meniscal repair and replacement techniques, allowing for effective translation to clinical applications.

Surgical repair of meniscal root avulsions has been traditionally limited to patients under the age of 35–40. However, there is a paradigm shift, with a contemporary inclination towards meniscus preservation through the utilization of repair techniques whenever feasible. The 2019 consensus of the European Society for Sports Traumatology, Knee Surgery and Arthroscopy (ESSKA) [42], drawing from a compilation of studies encompassing patients from 9 to 58 years of age, 'clearly states that numerous meniscus tears that were considered irreparable should be repaired, e.g., older tears'. This shift is substantiated by recent follow-up studies that report the advantages of extending the range for surgical intervention to over 60 years of age [41,51]. Hence, this matter was carefully considered, and we decided to expand the age range of the young group to 75 years old, aligning with the aforementioned findings and to conform to the criteria of the orthopedic surgeons who are part of our research team. We consider that menisci from donors who are older than 75 years of age are representative of a group of patients who are too advanced in age to be routinely considered eligible for meniscal root repair surgery.

The experimental design involved testing isolated meniscus–suture constructs instead of menisci attached to tibial bones. This setup is consistent with prior works on the same topic [29,37,38,52]. It was selected to specifically focus on the behavior of the tissue–suture interface, excluding factors dependent on tibial fixation. Also in line with previous studies [29,37,38,52], the suture was oriented with the loading direction, eliminating shear friction and concentrating the tensile effect on the meniscus–suture interface, which was our area of interest. This orientation of the meniscal horn aimed to maximize tissue damage, representing a conservative criterion for detecting the maximum load the repair could withstand.

A single simple suture was utilized at the meniscal horn, deviating from common surgical options. This choice facilitates computation of the forces acting on the tissue–suture interface. Multiple stitches distribute the pulling load among them, posing challenges for achieving uniform distribution under laboratory conditions. Despite neither being the most resistant nor the most common suture technique for meniscal root repair, the use of a simple stitch eases the quantification of the meniscus–thread contact area, essential for calculating stress at the suture–meniscus interface. This simplicity minimizes the impact of meniscus anatomy or surgeon performance compared to more complex suture techniques.

To accurately detect cut-out initiation, automatic detection using a previously validated method [46] was utilized and further verified by checking the recorded video images.

The results confirmed that, as formerly reported [46], tearing was initiated slightly before the first local maximum of the load–deformation curve. In the scientific literature, the maximum force borne by the specimen, F_u, has usually been selected as the parameter to characterize the failure of repaired menisci under suture-induced traction. The authors consider it more relevant to study the cut-out initiation point, i.e., F_c at the specimen level or S_c at the tissue level, although F_u has also been calculated for comparison purposes. F_u was found to be higher than F_c by nearly 30% for older human menisci and approximately 20% for the porcine group. After cut-out initiation, the progression of the tear may follow different paths within the meniscus of variable thickness. Consequently, the load–deformation curve exhibited a highly heterogeneous pattern from cut-out initiation to the absolute maximum force. This heterogeneity under laboratory conditions is expected to be more pronounced In a repaired knee subjected to physiological loads. Therefore, we believe that the cut-out initiation provides a better basis for comparing results between different groups because it is the starting point of the loss of structural integrity of the sample and it is less influenced by the post-tearing initiation trajectory.

Cryoprotectant solutions have been proposed to preserve biological tissues from damage caused by freezing, but we discarded their use for two main reasons. Primarily, they are not customarily used for preserving specimens in either of the two surrogate models of the human meniscus that our study focuses on. Additionally, there is a lack of data on their ability to maintain the mechanical properties of the suture meniscal horn better than the preservation method we used, particularly concerning elasticity and resistance.

Some studies have examined the resistance of meniscal horns to suture-induced traction, focusing on assessing various suturing techniques and devices for root reattachment. Antz et al. conducted a load-to-failure test on human menisci aged 46–64 years using two simple stitches with a #2 UHMWPE surgical suture on the posterior horn [37]. Despite differences in testing protocols, their reported values align with our findings for the young group, with a mean initial peak force of 137 $\pm$ 49 N for two stitches compared to our F_c of 60.13 $\pm$ 19.91 N with a single stitch. In accordance with our results, the initial peak force must be very close to our F_c. Vertullo et al. reported an ultimate load of 94.29 $\pm$ 7.99 N for a group of posterior medial menisci with a mean age of 54 $\pm$ 4 years using two simple #2 sutures [52]. This result is also compatible with our findings, considering F_u values of 75.08 $\pm$ 26.62 N for the young group and 70.49 $\pm$ 16.92 N for the older group, acknowledging the use of two stitches and prior sample weakening by 1000 cycles of under-critical load. It must be highlighted that in cases where more than one suture is employed, inevitable asymmetries in traction distribution may occur. Consequently, when the interface of one stitch reaches the tension threshold, initiating tearing, it is highly probable that the other stitches are at a lower load level. Mitchell et al. found a mean ultimate force of 58.2 $\pm$ 29.6 N when testing posterior horns of human medial menisci (ages 48 to 88 years) with one simple stitch [38]. The lower values observed may be attributed to their use of a #0 UHMWPE surgical suture, implying a thinner suture diameter and resistant area than in our study. Unfortunately, meniscus thickness was not reported, which does not permit us to estimate S_c in that study. Kopf et al. tested posterior human menisci aged 18–51 years in a load-to-failure test with two #2 single UHMWPE sutures, finding an ultimate load of 64.10 $\pm$ 22.50 N [53]. This value is considerably lower than other works and our younger group, factoring in the difference from the use of a single stitch.

Concerning meniscal horn elasticity, few studies have specifically addressed deformation at the sutured horn [29,46]. While some biomechanical works have reported on meniscus–suture construct stiffness, the results vary widely, even when restricted to UHMWPE threads. Stiffness for two simple sutures ranged from 774 $\pm$ 730 N/mm [38] using suture #0 to 24.55 $\pm$ 4.05 N/mm [52] with #2 sutures. Dispersion may be due to differences in suture thread lengths, which are often unreported, and varied cyclic loading histories before the load-to-failure test, which are known to influence the elasticity of surgical sutures if insufficient resting time is allowed [54].

Our results for the porcine model fall within the previously published range of failure loads. However, it is noteworthy that the range of published data is considerably wide. In a study where a meniscus was also sutured with a single simple stitch, utilizing a UHMWPE #2 suture [17], an ultimate load of 169 ± 73.4 N was reported. The result aligns with our outcome of 206.79 ± 74.31 N, considering that in that study, the specimens were pre-weakened with 1000 cycles of load between 5 and 20 N. In contrast, other tests involving porcine models yielded results for two sutures ranging from 129.8 ± 14.6 N [19] to 221.67 ± 43 N [29], although various suturing techniques and suture materials were employed. Beyond the variations attributed to different testing protocols, these disparities may be attributed to the diverse origin of the animal. While animal surrogate models provide the advantage of ensuring uniformity among donors, variations in breeds, types of feeding, stabling conditions, etc., can introduce significant differences between groups of diverse origins, even if they belong to the same species. This concern has been previously raised, and we believe it warrants further investigation [55].

In a study by Camarda et al., a comparison of the cyclic stiffness and ultimate load in load-to-failure tests on posterior horns of humans aged 50–70 years and porcine samples was undertaken [35]. Specimens sutured with three simple stitches using #2 UHMWPE threads were subjected to 1000 cycles of loading ranging from 10 N to 30 N, followed by a load-to-failure test. The study reports a 2.4-times-higher ultimate load for the porcine model compared to the human model, which is consistent with our findings of a 2.7-times-higher F_u for the porcine model compared to the younger human model, with no previous cyclic loading that may have weakened the tissue. Their human model also exhibited lower cyclic stiffness of the meniscus–suture construct in contrast to our results, which did not find any differences in deformation. However, the parameters are not comparable because our study was limited to the meniscus around the suture area, and we performed a direct load-to-failure test.

The mechanical properties of sutured meniscal tissue have not been studied so far. With the data extracted from this study, it becomes possible to fit the material properties of meniscal tissue of different origins in the area surrounding the suture hole, thus contributing to the development of computational models with more realistic behavior. On the other hand, the study highlights the need for further research to analyze the mechanical behavior of the sutured meniscal horn in situations where the viscoelastic nature of the meniscal tissue has a significant influence, such as in the case of dynamic impact loads or quasi-stationary loads maintained over a long period of time. Additionally, the authors maintain it will be interesting to conduct a thorough study of the alteration that the suture produces in the mechanical behavior of the meniscal root in the immediate postoperative period. Furthermore, in vivo studies could allow for an assessment of the progress of the mechanical properties of the sutured root as healing progresses, evaluating whether its natural meniscal properties can be fully restored.

This study is subjected to inherent limitations associated with in vitro testing, notably the absence of the healing effect observed over time. The use of an isolated meniscal setup implies a deliberate exclusion of the influences from surrounding soft tissues, ligaments, cartilage, and bones. The rationale for this decision has been discussed above. A single simple suture was employed, as also previously justified. Therefore, the cut-out and ultimate forces reported in the results section do not quantitatively reflect the maximum traction forces that the clinically repaired meniscus, using other more-common suturing techniques, might withstand. The load-to-failure test was conducted at 0.1 mm/s, representing a quasi-static displacement velocity lower than expected for knee movement in daily activities [54]. However, given the viscoelastic behavior of the meniscus reported in compression tests [55,56], further research is warranted to comprehensively understand the dynamic response of the sutured meniscal tissue.

This minimizes the potential influence of viscous effects on the test, a characteristic not addressed in this study. All the animals came from the same local slaughterhouse, which may raise questions about whether porcine models from other origins would give different results [57].

5. Conclusions

In in vitro experimental tests to analyze the mechanical behavior of the sutured horn of the adult human meniscus at an eligible age to be routinely repaired (below 75 years old), neither the porcine tissue nor the cadaveric tissue from donors of advanced age (over 75 years old) could provide accurate models. However, each one could be adequate for examining specific characteristics depending on the interest of the study; the older human meniscus model is a preferable surrogate model in studies on the resistance of sutured meniscal root specimens; conversely, porcine models provide a more accurate representation in works involving deformation-related properties of the meniscus around the suture site at the tissue level.

Author Contributions: Conceptualization, A.P.-T., A.P.-B. and M.P.-N.; methodology, A.P.-T., A.P.-B. and M.P.-N.; software, M.B.E.-C. and S.M.-V.; validation, A.P.-T., M.B.E.-C. and S.M.-V.; formal analysis, A.P.-T.; investigation, all authors; resources, all authors; data curation, all authors; writing original draft preparation, A.P.-T., A.P.-B. and M.P.-N.; writing—review and editing, all authors; visualization, S.M.-V. and M.B.E.-C.; supervision, A.P.-B. and M.P.-N.; project administration, A.P.-B. and M.P.-N.; funding acquisition, M.P.-N. All authors have read and agreed to the published version of the manuscript.

Funding: This research was funded by Junta de Andalucía, Spain, grant number UMA20-FEDERJA-116.

Institutional Review Board Statement: The study was conducted in accordance with the Declaration of Helsinki and approved (including the animal study protocol) by the Institutional Ethics Committee of the University of Málaga, Spain (Comité Ético de Experimentación de la Universidad de Málaga, CEUMA) (protocol code 69-2018-H 2 October 2018).

Informed Consent Statement: Patient consent was waived due to the knees being obtained from an authorized provider. The provider certified that the knees came from donors who voluntarily had donated their bodies to science without financial compensation, in compliance with the legislation of the country of origin, and who had previously signed the corresponding consent forms.

Data Availability Statement: The data presented in this study are available on request from the corresponding author. The data are not publicly available because they include the demographic data of the donors.

Acknowledgments: We thank A. Espejo-Baena and A. Espejo-Reina, orthopedic surgeons of the BIOCLINA group, for their invaluable collaboration.

Conflicts of Interest: The authors declare no conflicts of interest.

References

1. Fox, A.J.S.; Bedi, A.; Rodeo, S.A. The basic science of human knee menisci: Structure, composition, and function. *Sports Health* **2012**, *4*, 340–351. [CrossRef]
2. Perez-Blanca, A.; Espejo-Baena, A.; Amat Trujillo, D.; Prado Nóvoa, M.; Espejo-Reina, A.; Quintero López, C.; Ezquerro Juanco, F. Comparative Biomechanical Study on Contact Alterations after lateral meniscus posterior root avulsion, transosseous reinsertion, and total meniscectomy. *Arthroscopy* **2016**, *32*, 624–633. [CrossRef]
3. Allaire, R.; Muriuki, M.; Gilbertson, L.; Harner, C.D. Biomechanical consequences of a tear of the posterior root of the medial meniscus: Similar to total meniscectomy. *J. Bone Jt. Surg. Am.* **2008**, *90*, 1922–1931. [CrossRef]
4. Espejo-Reina, A.; Prado-Novoa, M.; Espejo-Baena, A.; Peña-Trabalon, A.; Perez-Blanca, A. Biomechanical consequences of anterior root detachment of the lateral meniscus and its reinsertion. *Sci. Rep.* **2022**, *12*, 6182. [CrossRef] [PubMed]
5. Shiwaku, K.; Kamiya, T.; Suzuki, D.; Yamakawa, S.; Otsubo, H.; Suzuki, T.; Takahashi, K.; Okada, Y.; Teramoto, A.; Ohnishi, H.; et al. The role of the medial meniscus in anterior knee stability. *Orthop. J. Sport Med.* **2022**, *10*, 23259671221132845. [CrossRef]
6. Musahl, V.; Citak, M.; O'Loughlin, P.; Choi, D.; Bedi, A.; Pearle, A. The effect of medial versus lateral meniscectomy on the stability of the anterior cruciate ligament-deficient knee. *Am. J. Sports Med.* **2010**, *38*, 1591–1597. [CrossRef] [PubMed]
7. Shoemaker, S.; Markolf, K. The role of the meniscus in the anterior-posterior stability of the loaded anterior cruciate-deficient knee. Effects of partial versus total excision. *J. Bone Jt. Surg. Am.* **1986**, *68*, 71–79. [CrossRef]
8. Krych, A.; Reardon, P.; Johnson, N.; Mohan, R.; Peter, L.; Levy, B.; Stuart, M.J. Non-operative management of medial meniscus posterior horn root tears is associated with worsening arthritis and poor clinical outcome at 5-year follow-up. *Knee Surg. Sports Traumatol. Arthrosc.* **2017**, *25*, 383–389. [CrossRef]
9. Hussain, Z.B.; Chahla, J.; Mandelbaum, B.R.; Gomoll, A.H.; LaPrade, R.F. The role of meniscal tears in spontaneous osteonecrosis of the knee: A systematic review of suspected etiology and a call to revisit nomenclature. *Am. J. Sports Med.* **2019**, *47*, 501–507. [CrossRef]

10. Pache, S.; Aman, Z.S.; Kennedy, M.; Nakama, G.Y.; Moatshe, G.; Ziegler, C.; LaPrade, R.F. Meniscal root tears: Current concepts review. *Arch. Bone Jt. Surg.* **2018**, *6*, 250–259.
11. Ahn, J.H.; Wang, J.H.; Yoo, J.C.; Noh, H.K.; Park, J.H. A pull out suture for transection of the posterior horn of the medial meniscus: Using a posterior trans-septal portal. *Knee Surg. Sports Traumatol. Arthrosc.* **2007**, *15*, 1510–1513. [CrossRef] [PubMed]
12. Feucht, M.J.; Grande, E.; Brunhuber, J.; Burgkart, R.; Imhoff, A.B.; Braun, S. Biomechanical evaluation of different suture techniques for arthroscopic transtibial pull-out repair of posterior medial meniscus root tears. *Am. J. Sports Med.* **2013**, *41*, 2784–2790. [CrossRef]
13. Balke, M.; Akoto, R.; Offerhaus, C.; Hoeher, J. Suture anchor refixation of meniscal root tears without an additional portal. *Arthrosc. Tech.* **2018**, *7*, e511–e515. [CrossRef] [PubMed]
14. Espejo-Reina, A.; Prado-Novoa, M.; Espejo-Reina, M.J.; Gómez-Cáceres, A.; Dalla Rosa-Nogales, J.; Espejo-Baena, A. Non anatomic reinsertion after amputation of the anterior horn of the lateral meniscus. *Orthop. Traumatol. Surg. Res.* **2019**, *105*, 1115–1118. [CrossRef]
15. Robinson, J.R.; Frank, E.G.; Hunter, A.J.; Jermin, P.J.; Gill, H.S. The strength of transosseous medial meniscal root repair using a simple suture technique is dependent on suture material and position. *Am. J. Sports Med.* **2018**, *46*, 924–932. [CrossRef]
16. Wu, J.L.; Lee, C.H.; Yang, C.T.; Chang, C.M.; Li, G.; Cheng, C.K.; Chen, C.H.; Huang, H.S.; Lai, Y.S. Novel technique for repairing posterior medial meniscus root tears using porcine knees and biomechanical study. *PLoS ONE* **2018**, *13*, e0192027. [CrossRef]
17. Feucht, M.J.; Grande, E.; Brunhuber, J.; Rosenstiel, N.; Burgkart, R.; Imhoff, A.B.; Braun, S. Biomechanical evaluation of different suture materials for arthroscopic transtibial pull-out repair of posterior meniscus root tears. *Knee Surg. Sports Traumatol. Arthrosc.* **2015**, *23*, 132–139. [CrossRef]
18. Chung, K.S.; Choi, C.H.; Bae, T.S.; Ha, J.K.; Jun, D.J.; Wang, J.H.; Kim, J.G. Comparison of tibiofemoral contact mechanics after various transtibial and all-inside fixation techniques for medial meniscus posterior root radial tears in a porcine model. *Arthroscopy* **2018**, *34*, 1060–1068. [CrossRef]
19. Fujii, M.; Furumatsu, T.; Xue, H.; Miyazawa, S.; Kodama, Y.; Hino, T.; Kamatsuki, Y.; Ozaki, T. Tensile strength of the pullout repair technique for the medial meniscus posterior root tear: A porcine study. *Int. Orthop.* **2017**, *41*, 2113–2118. [CrossRef] [PubMed]
20. Rosslenbroich, S.B.; Borgmann, J.; Herbort, M.; Raschke, M.J.; Petersen, W.; Zantop, T. Root tear of the meniscus: Biomechanical evaluation of an arthroscopic refixation technique. *Arch. Orthop. Trauma Surg.* **2013**, *133*, 111–115. [CrossRef]
21. Okimura, S.; Mae, T.; Tachibana, Y.; Iuchi, R.; Nakata, K.; Yamashita, T.; Shino, K. Biomechanical comparison of meniscus-suture constructs for pullout repair of medial meniscus posterior root tears. *J. Exp. Orthop.* **2019**, *6*, 17. [CrossRef] [PubMed]
22. Forkel, P.; Foehr, P.; Meyer, J.C.; Herbst, E.; Petersen, W.; Brucker, P.U.; Burgkart, R.; Imhoff, A.B. Biomechanical and viscoelastic properties of different posterior meniscal root fixation techniques. *Knee Surg. Sports Traumatol. Arthrosc.* **2017**, *25*, 403–410. [CrossRef] [PubMed]
23. Wu, S.-H.; Yeh, T.-T.; Hsu, W.-C.; Wu, A.T.; Li, G.; Chen, C.-H.; Lee, C.-H.; Wu, J.-L. Biomechanical comparison of four tibial fixation techniques for meniscal root sutures in posterior medial meniscus root repair: A porcine study. *J. Orthop. Transl.* **2020**, *24*, 144–149. [CrossRef]
24. Prado-Novoa, M.; Perez-Blanca, A.; Espejo-Reina, A.; Espejo-Reina, M.J.; Espejo-Baena, A. Initial biomechanical properties of transtibial meniscal root repair are improved by using a knotless anchor as a post-insertion tensioning device. *Sci. Rep.* **2020**, *10*, 1748. [CrossRef]
25. Feucht, M.J.; Grande, E.; Brunhuber, J.; Rosenstiel, N.; Burgkart, R.; Imhoff, A.B.; Braun, S. Biomechanical comparison between suture anchor and transtibial pull-out repair for posterior medial meniscus root tears. *Am. J. Sports Med.* **2014**, *42*, 187–193. [CrossRef]
26. Kim, Y.M.; Joo, Y.B.; Noh, C.K.; Park, I.Y. The optimal suture site for the repair of posterior horn root tears: Biomechanical evaluation of pullout strength in porcine menisci. *Knee Surg. Relat. Res.* **2016**, *28*, 147–152. [CrossRef] [PubMed]
27. Seo, J.H.; Li, G.; Shetty, G.M.; Kim, J.H.; Bae, J.H.; Jo, M.L.; Kim, J.S.; Lee, S.J.; Nha, K.W. Effect of Repair of Radial Tears at the Root of the Posterior Horn of the Medial meniscus with the pullout suture technique: A biomechanical study using porcine knees. *Arthroscopy* **2009**, *25*, 1281–1287. [CrossRef]
28. Stärke, C.; Kopf, S.; Gröbel, K.H.; Becker, R. The Effect of a Nonanatomic repair of the meniscal horn attachment on meniscal tension: A biomechanical study. *Arthroscopy* **2010**, *26*, 358–365. [CrossRef]
29. Perez-Blanca, A.; Prado Nóvoa, M.; Lombardo Torre, M.; Espejo-Reina, A.; Ezquerro Juanco, F.; Espejo-Baena, A. The role of suture cutout in the failure of meniscal root repair during the early post-operative period: A biomechanical study. *Int. Orthop.* **2018**, *42*, 811–818. [CrossRef]
30. Takroni, T.; Laouar, L.; Adesida, A.; Elliott, J.A.W.; Jomha, N.M. Anatomical study: Comparing the human, sheep and pig knee meniscus. *J. Exp. Orthop.* **2016**, *3*, 35. [CrossRef]
31. Deponti, D.; Di Giancamillo, A.; Scotti, C.; Peretti, G.M.; Martin, I. Animal models for meniscus repair and regeneration. *J. Tissue Eng. Regen. Med.* **2015**, *9*, 512–527. [CrossRef] [PubMed]
32. Kyle Martin, R.; Gillis, D.; Leiter, J.; Shantz, J.S.; MacDonald, P. A porcine knee model is valid for use in the evaluation of arthroscopic skills: A pilot study. *Clin. Orthop. Relat. Res.* **2016**, *474*, 965–970. [CrossRef] [PubMed]
33. Sweigart, M.A.; Zhu, C.F.; Burt, D.M.; Deholl, P.D.; Agrawal, C.M.; Clanton, T.O.; Athanasiou, K.A. Intraspecies and interspecies comparison of the compressive properties of the medial meniscus. *Ann. Biomed. Eng.* **2004**, *32*, 1569–1579. [CrossRef]
34. Polito, U.; Andreis, M.; Di Giancamillo, A.; Modina, S.; Scurati, R.; Marmotti, A.; Michielon, G.; Domenicucci, M.; Lombardo, M.; Di Giancamillo, M.; et al. Clinical anatomy of the meniscus in animal models: Pros and cons. *J. Biol. Regul. Homeost. Agents* **2020**, *34*, 197.
35. Camarda, L.; Bologna, E.; Pavan, D.; Morello, F.; Monachino, F.; Giacco, F.; Zingales, M. Posterior meniscal root repair: A biomechanical comparison between human and porcine menisci. *Muscle Ligaments Tendons J.* **2019**, *9*, 76–81. [CrossRef]

36. Espejo-Reina, A.; Prado-Novoa, M.; Espejo-Baena, A.; Estebanez, B.; Perez-Blanca, A. Improved tibiofemoral contact restoration after transtibial reinsertion of the anterior root of the lateral meniscus compared to in situ repair: A biomechanical study. *Int. Orthop.* **2023**, *47*, 2419–2427. [CrossRef]
37. Anz, A.W.; Branch, E.A.; Saliman, J.D. Biomechanical comparison of arthroscopic repair constructs for meniscal root tears. *Am. J. Sports Med.* **2014**, *42*, 2699–2706. [CrossRef]
38. Mitchell, R.; Pitts, R.; Kim, Y.M.; Matava, M.J. Medial meniscal root avulsion: A biomechanical comparison of 4 different repair constructs. *Arthroscopy* **2016**, *32*, 111–119. [CrossRef]
39. Krych, A.J.; Johnson, N.R.; Wu, I.T.; Smith, P.A.; Stuart, M.J. A simple cinch is superior to a locking loop for meniscus root repair: A human biomechanical comparison of suture constructs in a transtibial pull-out model. *Knee Surg. Sports Traumatol. Arthrosc.* **2018**, *26*, 2239–2244. [CrossRef] [PubMed]
40. Poland, S.; Everhart, J.S.; Kim, W.; Axcell, K.; Magnussen, R.A.; Flanigan, D.C. Age of 40 years or older does not affect meniscal repair failure risk at 5 years. *Arthroscopy* **2019**, *35*, 1527–1532. [CrossRef]
41. Ventura, M.; Seabra, P.; Oliveira, J.; Sousa, P.; Quesado, M.; Sousa, H.; Pereira, R.; Costa, A.; Carvalho, P. Meniscal Injuries in Patients Aged 40 Years or Older: A comparative study between meniscal repair and partial meniscectomy. *Cureus* **2023**, *15*, e33270. [CrossRef] [PubMed]
42. Kopf, S.; Beaufils, P.; Hirschmann, M.T.; Rotigliano, N.; Ollivier, M.; Pereira, H.; Verdonk, R.; Darabos, N.; Ntagiopoulos, P.; Dejour, D.; et al. Management of traumatic meniscus tears: The 2019 ESSKA meniscus consensus. *Knee Surg. Sports Traumatol. Arthrosc.* **2020**, *28*, 1177–1194. [CrossRef] [PubMed]
43. Pauli, C.; Grogan, S.; Patil, S.; Otsuki, S.; Hasegawa, A.; Koziol, J.; Lotz, M.; D'Lima, D.D. Macroscopic and histopathologic analysis of human knee menisci in aging and osteoarthritis. *Osteoarthr. Cartil.* **2011**, *19*, 1132–1141. [CrossRef] [PubMed]
44. Kim, Y.M.; Rhee, K.J.; Lee, J.K.; Hwang, D.S.; Yang, J.Y.; Kim, S.J. Arthroscopic pullout repair of a complete radial tear of the tibial attachment site of the medial meniscus posterior horn. *Arthroscopy* **2006**, *22*, 795.e1–795.e4. [CrossRef] [PubMed]
45. Menge, T.J.; Chahla, J.; Dean, C.S.; Mitchell, J.J.; Moatshe, G.; LaPrade, R.F. Anterior meniscal root repair using a transtibial double-tunnel pullout technique. *Arthrosc. Tech.* **2016**, *5*, e679–e684. [CrossRef] [PubMed]
46. Pérez de la Blanca, A. Biomecánica de la Reinserción Transtibial de la raíz Posterior del Menisco Lateral de la Rodilla: Avances en la Técnica de Reparación. Ph.D. Thesis, University of Málaga, Málaga, Spain, 2019.
47. Cerminara, A.J.; LaPrade, C.M.; Smith, S.D.; Ellman, M.B.; Wijdicks, C.A.; LaPrade, R.F. Biomechanical evaluation of a transtibial pull-out meniscal root repair challenging the bungee effect. *Am. J. Sports Med.* **2014**, *42*, 2988–2995. [CrossRef]
48. Faul, F.; Erdfelder, E.; Lang, A.G.; Buchner, A. G*Power 3: A flexible statistical power analysis program for the social, behavioral, and biomedical sciences. *Behav. Res. Methods.* **2007**, *39*, 175–191. [CrossRef] [PubMed]
49. Peña-Trabalon, A.; Perez-Blanca, A.; Moreno-Vegas, S.; Estebanez Campos, M.B.; Prado-Novoa, M. Age influence on resistance and deformation of the human sutured meniscal horn in the immediate postoperative period. *Front. Bioeng. Biotechnol.* **2024**, *11*, 1249982. [CrossRef]
50. Bansal, S.; Keah, N.M.; Neuwirth, A.L.; O'Reilly, O.; Qu, F.; Seiber, B.N.; Mandalapu, S.; Mauck, R.L.; Zgonis, M.H. large animal models of meniscus repair and regeneration: A systematic review of the state of the field. *Tissue Eng. Part C Methods* **2017**, *23*, 661. [CrossRef]
51. Husen, M.; Kennedy, N.I.; Till, S.; Reinholz, A.; Stuart, M.J.; Krych, A.J.; Saris, D.B. Benefits of meniscal repair in selected patients aged 60 years and older. *Orthop. J. Sport Med.* **2022**, *10*, 23259671221117492. [CrossRef]
52. Vertullo, C.J.; Cadman, J.; Dabirrahmani, D.; Appleyard, R. Biomechanical comparison of an all-inside meniscal repair device construct versus pullout sutures for arthroscopic transtibial repair of posterior medial meniscus root tears a matched-pair cadaveric study. *Orthop. J. Sport Med.* **2021**, *9*, 23259671211000464. [CrossRef] [PubMed]
53. Kopf, S.; Colvin, A.C.; Muriuki, M.; Zhang, X.; Harner, C.D. Meniscal root suturing techniques implications for root fixation. *Am. J. Sports Med.* **2011**, *39*, 2141–2146. [CrossRef] [PubMed]
54. Prado-Novoa, M.; Perez-Sanchez, L.; Estebanez, B.; Moreno-Vegas, S.; Perez-Blanca, A. influence of loading conditions on the mechanical performance of multifilament coreless UHMWPE sutures used in orthopaedic surgery. *Materials* **2022**, *15*, 2573. [CrossRef] [PubMed]
55. Norberg, C.; Filippone, G.; Andreopoulos, F.; Best, T.M.; Baraga, M.; Jackson, A.R.; Travascio, F. Viscoelastic and equilibrium shear properties of human meniscus: Relationships with tissue structure and composition. *J. Biomech.* **2021**, *120*, 110343. [CrossRef]
56. Morejon, A.; Mantero, A.M.A.; Best, T.M.; Jackson, A.R.; Travascio, F. Mechanisms of energy dissipation and relationship with tissue composition in human meniscus. *Osteoarthr. Cartil.* **2022**, *30*, 605–612. [CrossRef]
57. de Roy, L.; Morejon, A.; Jackson, A.R.; Travascio, F.; Seitz, A.M. A comparative study on the biomechanical properties of porcine meniscus in confined compression. In Proceedings of the 28th Congress of the European Society of Biomechanics, Maastricht, The Netherlands, 9–12 July 2023.

applied sciences

MDPI

Article

Mechanical Characterization of the Male Lower Urinary Tract: Comparison among Soft Tissues from the Same Human Case Study

Alice Berardo [1,2,3,†], Maria Vittoria Mascolini [4,†], Chiara Giulia Fontanella [2,4,*], Martina Contran [5], Martina Todesco [1], Andrea Porzionato [5,6,7], Veronica Macchi [5,6,7], Raffaele De Caro [5,6,7], Rafael Boscolo-Berto [5,6,7,‡] and Emanuele Luigi Carniel [2,4,‡]

1. Department of Civil, Environmental and Architectural Engineering, University of Padova, 35131 Padova, Italy; alice.berardo@unipd.it (A.B.); martina.todesco@unipd.it (M.T.)
2. Centre for Mechanics of Biological Materials, University of Padova, 35131 Padova, Italy; emanueleluigi.carniel@unipd.it
3. Department of Biomedical Sciences, University of Padova, 35131 Padova, Italy
4. Department of Industrial Engineering, University of Padova, 35131 Padova, Italy; mariavittoria.mascolini@phd.unipd.it
5. Department of Neuroscience, Institute of Human Anatomy, University of Padova, 35121 Padova, Italy; martina.contran@unipd.it (M.C.); andrea.porzionato@unipd.it (A.P.); veronica.macchi@unipd.it (V.M.); raffaele.decaro@unipd.it (R.D.C.); rafael.boscoloberto@unipd.it (R.B.-B.)
6. Veneto Region Reference Centre for the Preservation and Use of Gifted Bodies, Veneto Region, 35100 Padua, Italy
7. National Reference Centre for the Preservation and Use of Gifted Bodies, 35121 Padua, Italy

* Correspondence: chiaragiulia.fontanella@unipd.it

† These authors contributed equally to this work.

‡ These authors are both last authors of the work.

Featured Application: An exhaustive mechanical characterization of the lower urinary tract represents a key point for the development of effective in silico models to provide additional information on dysfunctions and validate innovative devices for the treatment of related pathologies.

Abstract: Background: Nowadays, a challenging task concerns the biomechanical study of the human lower urinary tract (LUT) due to the variety of its tissues and the low availability of samples. Methods: This work attempted to further extend the knowledge through a comprehensive mechanical characterization of the male LUT by considering numerous tissues harvested from the same cadaver, including some never studied before. Samples of the bladder, urethra, prostate, Buck's fascia and tunica albuginea related to corpora cavernosa were considered and distinguished according to testing direction, specimen conformation and anatomical region. Uniaxial tensile and indentation tests were performed and ad hoc protocols were developed. Results: The tissues showed a non-linear and viscoelastic response but different mechanical properties due to their specific functionality and microstructural configuration. Tunica albuginea longitudinally displayed the highest stiffness (12.77 MPa), while the prostate transversally had the lowest one (0.66 MPa). The minimum stress relaxation degree (65.74%) was reached by the tunica albuginea and the maximum (88.55%) by the bladder. The prostate elastic modulus was shown to vary according to the presence of pathological changes at the microstructure. Conclusions: This is the first experimental work that considers the mechanical evaluation of the LUT tissues in relation to the same subject, setting the basis for future developments by expanding the sample population and for the development of effective in silico models to improve the solutions for most LUT pathologies.

Keywords: lower urinary tract; human soft tissue; urethral tissues; mechanical tests; prostate; bladder; tunica albuginea; Buck's fascia

1. Introduction

The human male lower urinary tract (LUT) consists of the bladder and the urethra, which in combination allow the storage and the voiding of urine. The bladder is the distensible and hollow organ acting as the reservoir of urine and in continuous connection to the upper urinary tract through the two ureters and to the urethra through its lowest part, the neck [1]. The urethra is the duct through which the urine passes and it is distinct along its length, from the bladder neck to the external urethral orifice, in five segments (pre-prostatic, prostatic, membranous, bulbar and penile). In males, the LUT is also completed by other tissues, such as the prostate, the Buck's fascia and the tunica albuginea, that are anatomically and functionally related to the main components. The prostate, immediately below the bladder neck, is an inverted pyramid-shaped gland surrounding the urethra in the prostatic region. In the distal region, the urethra passes through the corpus spongiosum, which composes the penis together with the two corpora cavernosa [2]. The Buck's fascia surrounds as a whole the corpus spongiosum and corpora cavernosa, each of which is enveloped by its own layer of tunica albuginea, which is thinner for the corpus spongiosum.

All these LUT components are soft connective tissues characterized by specific microstructural configurations determining their mechanical properties. The mechanical behavior of many human tissues has already been extensively investigated [3], but the research in LUT biomechanics is still at an early stage, mainly due to the heterogeneity of the tissues involved and the difficulty in the availability of samples. To date, few works in literature have analyzed the mechanics of these tissues.

Regarding the urethral tissue, in 2018 Masri et al. [4] mechanically characterized five human urethrae from male cadavers by separating membranous and spongy urethra portions and then testing the samples along the longitudinal and circumferential directions, while in 2020, Cunnane et al. [5] performed mechanical tests on human male urethrae obtained from patients that underwent gender reassignment surgery. The results revealed a nonlinear response with stiffening at higher pressures during both static and dynamic loadings and pronounced viscoelastic behavior with increasing stretch. In addition, Cunnane et al. also performed tensile tests by considering ring portions of the bulbar and penile urethra from the distal and proximal regions [5]. No difference between the directional responses at any stress was observed as well as between regional responses during circumferential or longitudinal extension and in terms of the relaxation or creep parameters defining the response in the two directions.

Lalla et al. [6] investigated the biomechanical properties in a rabbit model to test hypospadias operations by stretching ring specimens until rupture. In the same context of the failure tensile test, in 2021 Cunnane et al. [7] mechanically characterized 18 porcine urethral tissue subjected to different cryopreservation protocols to identify the one that best preserved the native tissue properties. The findings showed that urethral tissue should be stored at -20 °C, while the storage at -80 °C resulted in significantly altering the mechanical properties, failure mechanics and geometry of the tissue. Finally, Natali et al. [8] selected the equine urethra as a suitable animal model due to the similarity in terms of the lumen, histo-morphometric conformation and functional process of micturition to the human urethra. Tensile, stress–relaxation and inflation tests were performed. A significant difference in terms of stiffness resulted when the longitudinal proximal specimens were compared to all the other samples.

As concerns the mechanical investigation of the bladder tissue, in 2012, Zanetti et al. [9] studied the effect of strain rate, orientation and loading history on the behavior of porcine bladder tissue. In 2015 Natali et al. [10] performed uniaxial cyclic tests on porcine samples at different strain rates (10, 50, 100, 200, 500% s^{-1}). Transverse specimens showed greater stiffness and influence of the strain rate than those exhibited by apex-to-base ones.

Concerning prostate tissue, most of the biomechanical research has been carried out to study the changes in the mechanical properties of pathological conditions. Krouskop et al. [11] considered 113 specimens from human normal, benign prostatic hyperplasia (BPH) and cancer prostate tissues, which were cut into slab-shaped samples and tested under indentation, revealing that tissue from prostate with BPH resulted significantly softer than normal tissue, while cancers significantly stiffer than normal tissue at both strain ranges. Phipps et al. [12] collected 85 fresh chippings via transurethral resection of the prostate (TURP) from 17 patients with an ad hoc system for mechanical testing through an electromechanical shaker inducing a cyclic compressive load at actuation frequencies between 5 and 50 Hz to derive strong correlations between mechanical properties and morphology of the prostatic tissue, which may allow a more accurate prediction of the response to a given therapy [12]. Other indentation tests were also performed by Ahn et al. [13] and Kim et al. [14], while Ma et al. [15] tested prostate samples, cut parallel and perpendicular to the urethra, through uniaxial load–unload mechanical testing. Both healthy (cadaveric) as well as periurethral tissues obtained from radical prostatectomy were analyzed. These latter reported a wider range of tangent modulus with greater stiffness directly correlated with higher collagen content due to fibrosis and with moderate/severe LUT symptoms.

The mechanical basis of the tunica albuginea was only partially studied in human health conditions [16] or disease [17], and in an animal model [18]. In 1990, Bitsch et al. [16] performed tensile tests up to failure by adopting a tensiometer and exposing the samples to a stepwise increasing pressure. In 2022, Brady et al. [17] designed a testing strategy to characterize the mechanical properties of the tunica albuginea affected by Peyronie's Disease (PD), a progressive fibrosis that can cause tissue mineralization. Uniaxial tensile tests up to failure showed an altered tissue behavior with mineralization, supported by the literature studies on the atherosclerotic plaque mechanics which found higher moduli [19–21] in calcified plaques compared to healthy tissue.

No published studies investigated the biomechanical behavior of Buck's fascia or studied the overall mechanical properties of the different LUT tissues from the same subject. This ongoing challenge is attracting considerable interest among the scientific community due to the close relationship between the functionality of LUT constituent tissues and their mechanical performances. Assessing the mechanical response and identifying and validating the constitutive models of the entire LUT are fundamental steps to develop in silico models for surgical procedures and prosthesis design in order to propose alternative and innovative solutions for most pathologies, such as urinary incontinence and BPH, that nowadays are not always optimized or effective.

For the first time, this study aimed to simultaneously mechanically characterize the soft tissues composing the entire human male LUT, thanks to the combination of multiple mechanical tests (uniaxial tensile tests up to failure, stress–relaxation tests and indentation tests) on five different tissues, cut along different directions when possible. Bladder, urethra, prostate, Buck's fascia and tunica albuginea related to corpora cavernosa were obtained from a fresh-frozen 77-year-old male cadaver donor, according to the Body Donation Program of the Institute of Anatomy of the University of Padua [22].

Quantification and evaluation of the mechanical failure and time-dependent properties were performed, analyzing similarities and differences between tissues according to their functionality and in relation to the same subject. In addition, histological analyses were carried out to confirm the obtained results. This work could represent a turning point in advancing the development of effective in silico models for studying LUT dysfunctions. It could also contribute to the design of devices for treating LUT pathologies, the creation of reliable tissue response-mimicking phantoms for surgical planning, and advancements in tissue engineering. These applications extend from anatomical surgical reconstruction to addressing functional defects in the LUT.

2. Materials and Methods

2.1. Tissues Harvesting

In accordance with the Body Donation Program of the Institute of Anatomy of the University of Padua [22], bladder, urethra, prostate, Buck's fascia and tunica albuginea related to corpora cavernosa were obtained, without apparent anomalies, from a fresh-frozen 77-year-old male cadaver. His clinical records reported a remote TURP without significant complications or sequelae that could have affected the LUT tissues and functions. The gross in situ dissection and the following microscopic examination of the harvested tissue were consistent with the donor's medical history. The urethra and prostate were acquired entirely as structures, while other tissues were harvested as patches. Tissues were stored at −20 °C and thawed before testing.

2.2. Sample Preparation

All tissues were accurately prepared for the mechanical testing, according to the typology of the test and the nature and origin of the tissue (Figure 1). To perform the uniaxial tensile tests, rectangular strip or ring samples were used.

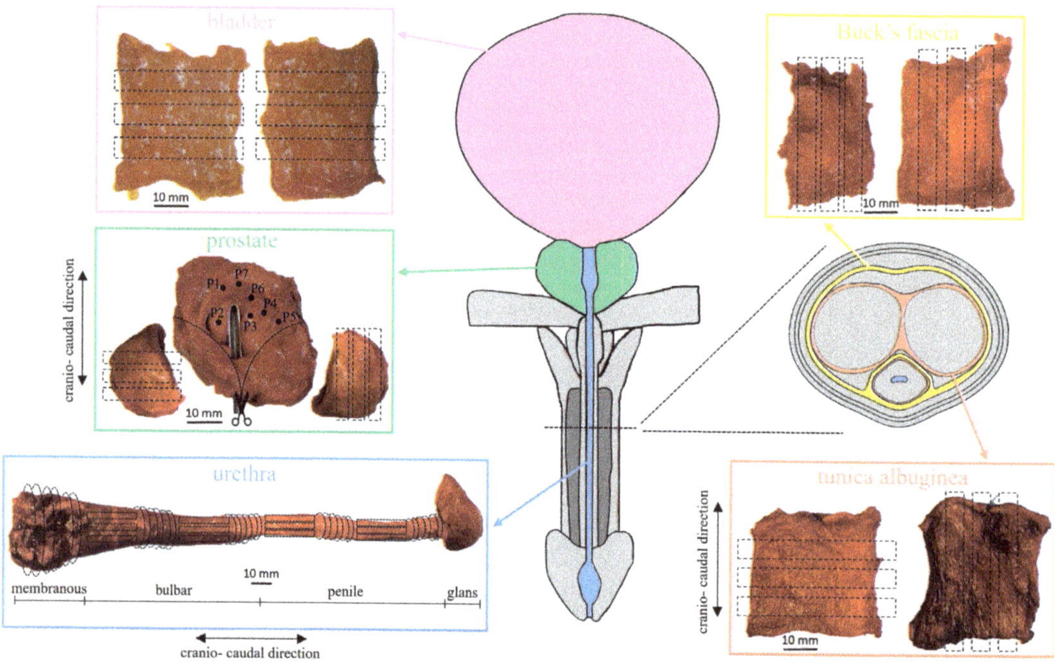

Figure 1. Anatomical illustration of the human male LUT and cross-section of the penis showing the relationship between bladder, urethra, prostate, Buck's fascia and tunica albuginea. For each acquired tissue, details for sample preparation are reported.

Bladder patches were cut in rectangular strips amounting to 6 samples (as reported in Table 1) with mean ± standard deviation (SD) width of 4.47 ± 0.67 mm and length of 19.71 ± 0.73 mm, as well as for Buck's fascia (width 2.17 ± 0.73 mm, length 20.00 ± 0.00 mm). In these cases, only one cut direction was considered because the specific harvesting site was unknown for the bladder, while no references were found in literature about Buck's fascia biomechanics since it appears a quite new field of research.

Secondly, tunica albuginea of the corpora cavernosa was tested along longitudinal and transversal directions with respect to the urethra, thus 5 rectangular strips (Table 1) were obtained from its native patch (width 3.44 ± 0.94 mm, length 19.82 ± 0.60 mm). The anisotropy was also investigated for the urethra and the prostate, according to the literature [4,5,8,15]. The extracted urethra, from the proximal to the distal end, included the membranous, bulbar and penile portions (Figure 1). For this reason, it was firstly transversally sectioned to distinguish the 3 specific portions. The membranous portion was again transversally cut obtaining 3 rings samples (Table 1) with a mean width of 4.31 ± 0.46 mm. On the other hand, both bulbar and penile portions were transversally sectioned in the middle to distinguish the distal and the proximal regions. Successively, each region of both urethral portions was transversally dissected obtaining 3 rings with a mean width of 4.52 ± 0.71 mm and a remaining wider ring, which was incised longitudinally and straightened. From this latter, 4 rectangular strips with an average width of 6.81 ± 1.47 mm and length of 14.53 ± 8.35 mm were obtained for each region. In total, 8 longitudinal and 6 transversal samples were gathered for both the bulbar and the penile urethral portions (Table 1).

Table 1. For each type of test and kind of tissue, the number of samples is reported together with the directions studied.

Test	Type	Tissue	No. of Samples	Direction
Uniaxial tensile	Failure	Bladder	6	-
		Buck's fascia	6	-
		Prostate	8	Longitudinal (#5) Transversal (#3)
		Tunica albuginea	5	Longitudinal (#3) Transversal (#2)
	Stress–relaxation	Bladder	6	-
		Tunica albuginea	3	Transversal
		Membranous urethra	3	Transversal
		Bulbar urethra	14	Longitudinal (#8) Transversal (#6)
		Penile urethra	14	Longitudinal (#8) Transversal (#6)
Indentation		Prostate	1	-

The acquired prostate was cut longitudinally in the apex region and the right and left lateral sides were isolated from the central portion. From the right side, 5 rectangular strips were trimmed along the direction longitudinal to the urethra, while 3 analogous samples, from the left side, were obtained along the transversal direction (width 8.01 ± 2.83 mm, length 16.25 ± 5.18 mm). On the remaining prostate part, indentation tests were performed. Both during and after the preparation procedure, a phosphate-buffered saline solution was used to keep all the samples moist so as not to alter their mechanical properties due to non-physiological dehydration.

Moreover, for the strip specimens, a further final step of preparation was performed by using a paper C-shaped support. Each end of the samples was interposed between one extremity of the support and a small paper strip, to which the Velcro was glued. The male sides of the Velcro in addition to a medical-grade cyanoacrylate glue (Glue Loctite 4013 Prism, Hisco, Houston, TX, USA) on their surface were used to fix the ends of the samples, facilitating their positioning within the grips of the mechanical tester and avoiding both the slippage and the misalignment of the specimens. In addition, the so prepared samples were clamped by adequate grips with textured pattern surfaces, and the closure pressure was adjusted to ensure the fixation but, at the same time, the absence of tissue damage. Afterward, the C-shaped support was cut along its length to allow the mechanical test. On the other hand, the ring specimens did not require any additional preparation, they were slid over the top and bottom posts of a T-shaped tensile fixture (post diameter 0.8 mm), specifically developed for the mechanical evaluation of small ring of material [23].

For all samples prepared for tensile tests, the grip-to-grip length and width were defined through the image processing software ImageJ (NIH, Bethesda, MD, USA, version 1.54d, 2020) while the thickness was measured by means of a manual caliper of 0.1 mm resolution at different positions along the length of the sample and the average value was finally considered. Mean ± SD values of width, thickness and length of the samples are reported in Table 2 for each tissue studied.

Table 2. For each tissue, mean ± SD values of width, thickness and length are reported in mm.

Tissue	Specimen	Width (Mean ± SD)	Thickness (Mean ± SD)	Length (Mean ± SD)
Bladder	Strip	4.47 ± 0.67	2.14 ± 0.31	19.71 ± 0.73
Buck's fascia	Strip	2.17 ± 0.73	0.87 ± 0.49	20.00 ± 0.00
Prostate	Strip	8.01 ± 2.83	1.96 ± 0.62	16.25 ± 5.18
Tunica albuginea	Strip	3.44 ± 0.94	2.38 ± 0.52	19.82 ± 0.60
Membranous urethra	Ring	4.31 ± 0.46	8.63 ± 2.85	14.34 ± 4.04
Bulbar urethra	Ring	4.76 ± 0.19	9.57 ± 1.56	14.00 ± 3.79
	Strip	7.12 ± 1.68	3.79 ± 0.99	21.45 ± 1.94
Penile urethra	Ring	4.29 ± 0.98	8.09 ± 0.91	13.83 ± 2.56
	Strip	6.47 ± 1.20	3.66 ± 0.51	22.87 ± 2.47

2.3. Experimental Protocols

Tests were performed at room temperature by means of the mechanical tester Biomomentum Mach-1 v500csst (Laval, QC, Canada) equipped with a 3-axis motion controller (resolution of 0.5 μm, max acceleration of 500 mm/s^2) and a uniaxial load cell of 250 N (accuracy of 0.0125 N). Experimental protocols for uniaxial tensile tests adopted for each tissue and indentation tests for the prostate tissue are reported below.

2.3.1. Uniaxial Tensile Test

Prior to the tensile tests, all the samples were subjected to a first preconditioning phase through the application of a sinusoidal displacement curve (consisting of 10 cycles), for a total applied strain of 5% and a curve frequency of 1.5 Hz. Then, the issue was allowed to recover during a resting phase of 30 s before starting the effective test.

Uniaxial tensile tests up to failure or stress–relaxation tests were performed. In detail, bladder, Buck's fascia, prostate and tunica albuginea strips were tested up to the failure by applying a progressive strain with a strain rate of 2.5% s^{-1} (as shown in Table 3). The tests were stopped when a decreasing force was measured or visible damage to the tissue was observed.

Table 3. Experimental protocol adopted according to the different biological tissues.

Test	Tissue	Strain Rate [%s^{-1}]	No. of Ramps	Strain Step [%]	Resting Time [s]
Failure	Bladder	2.5	-	-	-
	Buck's fascia	2.5	-	-	-
	Prostate	2.5	-	-	-
	Tunica albuginea	2.5	-	-	-
Stress–relaxation	Bladder	30	5	10	300
	Tunica albuginea	30	4	10	300
	Urethra	30	6	10	400

On the other hand, the stress–relaxation protocol included a defined number of consecutive almost instantaneous (strain rate of 30% s^{-1}) elongations, with each step of 10% strain and held constant for a fixed resting time. In particular, for the bladder 5 strain steps and a relaxation time of 300 s were considered, for the tunica albuginea 4 strain steps and 400 s of relaxation time were considered and for the urethra 6 strain steps and 400 s of relaxation time were considered (Table 3).

2.3.2. Indentation Test

The indentation test was performed only for the prostate, due to the limited sample availability for the other tissues. The central portion of the acquired prostate was placed on a flat base, to which the Velcro was glued to guarantee adherence and attachment to keep the sample fixed during the test. The anterior surface was measured by identifying 7 indentation sites which could represent the peculiarities and variability of the tissue

(Figure 1). The test was carried out by means of a spherical tip indenter (radius R of 3.18 mm) inducing an increasing indentation depth up to 4 mm at 0.2 mm s^{-1}. The indentations started once the load cell found contact with the tissue.

2.4. Histological Evaluations

Among the selected tissues, patch samples of the bladder and tunica albuginea related to corpora cavernosa were harvested with a dimension of 5 mm × 20 mm. Moreover, 5 mm thickness transversal sections of the whole prostate and penis were cut to obtain macrosections. All tissues were fixed for one week in formaldehyde 4% m/v, buffered according to Lille, pH 6.9 (100496, Sigma Aldrich, Darmstadt, Germany) in compliance with the solution-sample ratio of 20:1.

Tissue processing and paraffin embedding was carried out automatically (Leica TP1020, Leica Biosystems Nussloch GmbH, Nußloch, Germany) in the case of bladder and tunica albuginea patches, while manually for macroscopic sections of prostate and penis according to routine protocols. Thereafter, 5 μm thick sections were cut and stained with Hematoxylin-Eosin (W01030708, BioOptica Milano, Milan, Italy) and Azan-Mallory (04-001802, BioOptica Milano, Milan, Italy) for histological characterization. In both cases, all the sections were deparaffinized with xylene, hydrated through a decreasing scale of alcohol solutions, stained with kit-specific dyes, dehydrated through an increasing scale of alcohol solutions, clarified in xylene and mounted through Eukitt mounting medium (09-00250, BioOptica Milano, Milan, Italy).

2.5. Statistical Analysis

Median, 25th and 75th percentiles were calculated for all the tested samples by using Matlab (MATLAB and Statistics Toolbox Release 2022b, The MathWorks, Inc., Natick, MA, USA). No statistical tests were performed due to the limited number of tests per tissue.

3. Results

Mechanical tests were performed to study the different mechanical responses of the tissues composing the LUT, when possible examining the regional and directional tissue variability. From experimental tensile data, the strain was computed as the ratio between the elongation and the initial grip-to-grip length of the specimen, while the stress was computed as the ratio between the force measured by the load cell and the initial cross-sectional area of the sample.

Regarding the failure results, for all tested tissues the stress–strain curves were considered up to the inflection point assumed as the failure point of the tissue. At this point, the stress and strain values were extracted as respectively the ultimate tensile stress (UTS) and the strain at UTS to characterize the mechanical tensile response of the different tissues.

Median curves and confidence intervals for the tissues tested until failure are shown in Figure 2.

The elastic modulus E was extracted from the linear region of the stress–strain curves, which differs among the tissues. This region corresponded to specific strain ranges according to the tissue: for the bladder [19.42–29.99]%, for the Buck's fascia [13.05, 22.91]%, for the prostate [23.63–19.48]% along the longitudinal direction while [52.51–68.32]% along the transversal one and for the tunica albuginea [9.03, 19.04]% along the longitudinal direction and [10.52–19.02]% along the transversal one. E was then defined as the slope of these regions, computed by means of linear fitting ($R^2 > 0.99$), as performed in other studies [24,25].

Median values (25th, 75th percentiles) of the UTS, the strain at UTS and the elastic modulus are reported in Table 4. Among all, the highest stiffness of the elastic modulus was exhibited by the tunica albuginea along the transversal direction (median 40.44 MPa, [35.91, 44.97] MPa), followed by its longitudinal direction (12.77 MPa, [10.36, 17.70] MPa), while the lowest one by the prostate along the transversal direction (0.66 MPa, [0.63, 0.86] MPa). In accordance with these results, the tunica albuginea along the transversal direction showed also the highest UTS, while the minimum strain at UTS was recorded by its longitudinal direction. On the opposite, the prostate along the transversal direction exhibited the greater strain at break. Concerning the anisotropy of the prostate tissue, the elastic modulus, the UTS and the strain at UTS along the longitudinal direction resulted, in median terms, respectively, 2.34, 1.37 and 0.30 times the corresponding values along the transversal direction.

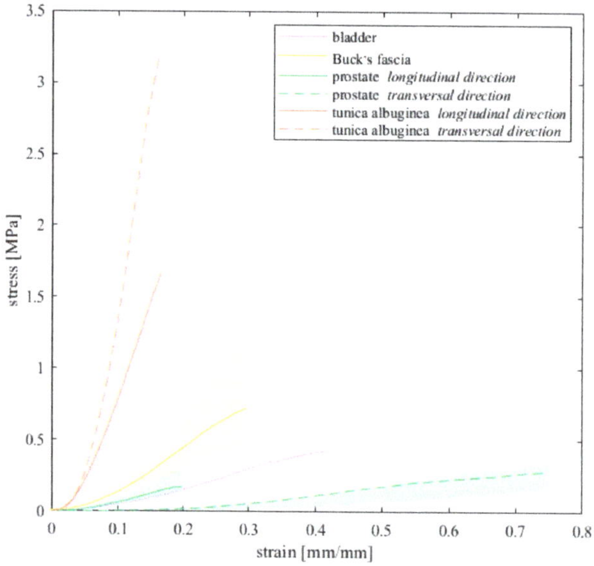

Figure 2. Stress–strain curves up to breakage resulted for all the tissues that were subjected to failure tests. Median curves are reported with 50% probability scatter bands.

Table 4. For each tissue tensioned up to failure, median values (25th, 75th percentiles) of the elastic modulus, UTS and strain at UTS are reported.

Tissue	*E* [MPa]	UTS [MPa]	Failure Strain [%]
Bladder	1.40 (1.16, 1.58)	0.47 (0.44, 0.54)	51.42 (47.23, 54.69)
Buck's fascia	3.86 (2.18, 7.46)	0.86 (0.66, 1.79)	47.79 (42.45, 50.73)
Prostate—L	1.54 (1.13, 2.35)	0.44 (0.26, 0.48)	25.37 (22.62, 45.38)
Prostate—T	0.66 (0.63, 0.86)	0.32 (0.29, 0.41)	85.45 (77.80, 124.19)
Tunica albuginea—L	12.77 (10.36, 17.70)	2.12 (1.39, 3.81)	21.14 (17.91, 26.97)
Tunica albuginea—T	40.44 (35.91, 44.97)	6.50 (3.64, 9.36)	24.27 (16.78, 31.75)

L = longitudinal, T = transversal.

Data from stress–relaxation tests were processed to analyze the response of each tested tissue during and after the relaxation phases. From stress–time data, the relaxation curves were extracted in response to each step strain and, to compare results at the same strain condition, normalized stress data were obtained as the ratio between stresses at the current time and at the beginning of the relaxation phase. Depending on the strain history assumed for each tissue, the normalized stress–time curves were separately determined and fitted by the following exponential formulations:

$$\sigma_{norm}(t) = 1 - \gamma_1 - \gamma_2 + \gamma_1\, e^{(-t/\tau 1)} + \gamma_2\, e^{(-t/\tau 2)} \quad (1)$$

Equation (1) describes the normalized stress σ_{norm} over time t and accounts for two viscous branches, whose parameters are γ_1 and γ_2 as relative stiffness and τ_1 and τ_2 as the time constants of relaxation. The number of branches was defined by a trade-off between the minimization of the number of the parameters and the adequate interpretation of the trend of the experimental data and agrees with previous studies on LUT tissues [8,9]. The quantity $\gamma_1 + \gamma_2$ accounts for the amount of stress relaxation of the normalized curves (in percentage it represents the degree of relaxation for each tissue). The resulting normalized median stress–time curves and the related 50% probability scatter bands are shown in Figure 3, while fitting parameters are reported in Table 5.

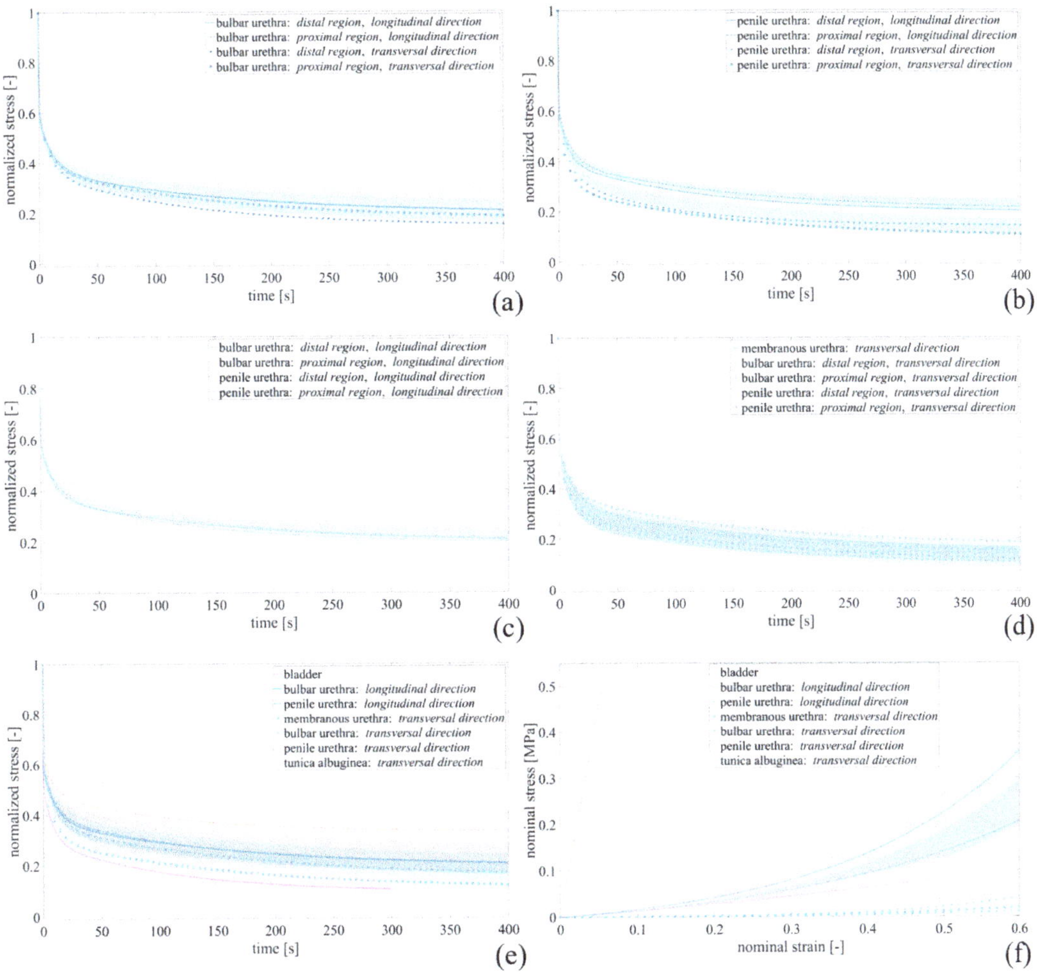

Figure 3. Relative stress–relaxation curves resulted for (**a**) bulbar and (**b**) penile urethra in distal and proximal regions, for all urethral samples tested along the (**c**) longitudinal and (**d**) transversal directions, and for (**e**) all tissues underwent stress–relaxation test. (**f**) Stress–strain curves obtained at equilibrium for bladder, tunica albuginea and urethra samples. Median curves are reported together with 50% probability scatter bands.

For the bulbar urethra (Figure 3a), differences in terms of relaxation degree were not found between distal (D) and proximal (P) regions along both the longitudinal (D 78.19%, [74.23, 82.94]%; P 78.12%, [67.24, 82.65]%) and the transversal (D 83.95%, [82.06, 86.26]%; P 80.68%, [76.14, 83.04]%) directions. Similarly, no differences resulted for the penile urethra (Figure 3b) between the D and P regions along the two tested directions.

Table 5. From the results of stress–relaxation tests, elastic (k, α) and viscous (γ_1, γ_2, τ_1, τ_2) parameters were obtained in terms of median values.

Tissue	k [kPa]	α	γ_1	γ_2	τ_1 [s]	τ_2 [s]
Bladder	120	1.20	0.63	0.26	1.01	65.39
Tunica albuginea—T	8640	2.05	0.46	0.20	1.03	68.91
Membranous urethra—T	5.40	4.94	0.63	0.24	1.21	96.26
Bulbar urethra—L	130	4.19	0.56	0.23	1.44	83.10
Bulbar urethra—T	3.00	5.13	0.58	0.25	1.49	104.80
Penile urethra—L	120	3.13	0.55	0.24	1.29	93.84
Penile urethra—T	4.70	6.94	0.60	0.26	1.22	72.56

L = longitudinal, T = transversal.

For these reasons, median values of the resulting parameters γ_1, γ_2, τ_1 and τ_2 are reported in Table 5 without distinguishing the distal and proximal regions.

Both the bulbar and penile urethral samples suggested different relaxation phases if tested along the longitudinal or the transversal direction. In particular, the penile urethra showed a stress decrease equal to 87.59% [83.31, 90.10]% along the transversal direction, while equal to 79.33% [75.63, 84.03]% along the longitudinal one.

Analyzing all the urethral samples tested longitudinally, the relaxation curves resulted independent of the harvesting region, as shown in Figure 3c. On the other hand, considering all the urethral samples tested transversally (Figure 3d), different urethral regions determined different degrees of relaxation, reporting the maximum value for the penile urethra (87.59%, [83.31, 90.10]%) and the minimum for the bulbar urethra (83.22%, [79.84, 86.02]%). All these curves obtained from urethral samples resulted in an intermediate region between those relating to the tunica albuginea tested along the transversal direction and those related to the bladder (Figure 3e) exhibited a final drop of normalized stress equal to 65.74% [53.47, 82.12]% and 88.55% [84.00, 91.18]%, respectively.

Moreover, original stress–time data were processed to analyze the tissue mechanical response at the end of relaxation phases, when all the time-dependent phenomena completely occurred. By considering the stress at the equilibrium point for each step strain applied for each specimen, non-linear stress (σ)–strain (ε) curves were separately obtained and fitted by the following exponential formulation:

$$\sigma\,(\varepsilon) = k/\alpha\,[e^{(\alpha\varepsilon)} - 1] \tag{2}$$

The parameter k specifies the tissues' initial stiffness, while α is related to the non-linearity of the response due to the stiffening. For each tissue, the resulting parameters (k, α) are reported in terms of median in Table 5. It can be observed that, among all the tested tissues, the tunica albuginea exhibited the highest stiffness with a value up to three scale orders above. Stress–strain curves obtained at equilibrium are reported in Figure 3f.

From the indentation test on the prostate sample, the reaction forces and the deformation were acquired for each indentation point.

Under the assumption of the tissue as a semi-infinite elastic body and considering that the indenter perfectly contacted with the tissue, E was fitted by means of the Hertz-Sneddon equation [13]:

$$F = 4/3\,E\,\left(1 - \nu^2\right)^{-1}\sqrt{R\delta^3} \tag{3}$$

where F is the maximum value of the reaction force, R is the indenter radius, δ is the indentation depth and ν is the Poisson's ratio, assumed equal to 0.499.

Figure 4 shows the mechanical responses resulting from the indentation points P1 and P5, placed in different regions of the prostate sample. In particular, P1 refers to one nodule formation on the prostate, while P5 refers to the non-altered tissue. From the experimental results, only the first part of the curve was fitted, for a maximum indentation depth of 1.5 mm and resulted in a good fit quality (reported in Table 6 by the R^2 parameter).

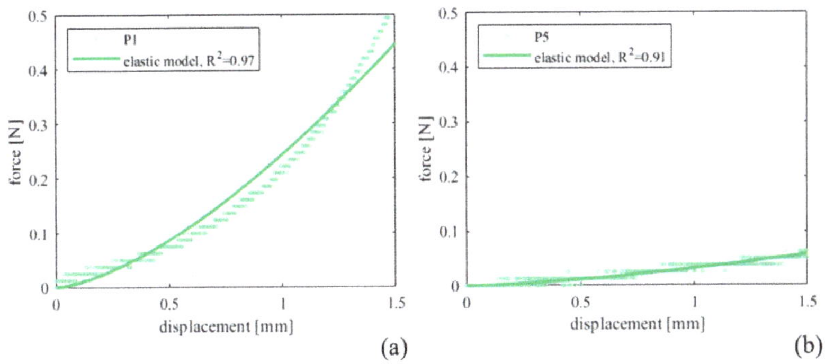

Figure 4. Indentation force–depth curves measured for (**a**) P1 and (**b**) P5 prostate indentation points and comparison with the fit performed by adopting Equation (3).

Table 6. Prostate elastic modulus in MPa (and R^2 in brackets) reported according to the indentation region (P1–P7).

P1	P2	P3	P4	P5	P6	P7
0.077 (0.97)	0.027 (0.98)	0.017 (0.46)	0.015 (0.85)	0.01 (0.91)	0.022 (0.90)	0.034 (0.98)

With reference to the histological analyses, Hematoxylin–Eosin staining was performed to observe the general layout of the tissues, as reported in Figure 5a,c for the prostate and bladder, respectively, and Figure 6 for the urethra, Buck's fascia and tunica albuginea. This specific staining highlights the prostate tubuloacinar glands surrounded by fibromuscular stroma (Figure 5a). Furthermore, Figure 6a shows both the urethral section as well as the corpora cavernosa. At the lower center (Figure 6b), the corpus spongiosum surrounding the bulbar urethra is clearly visible.

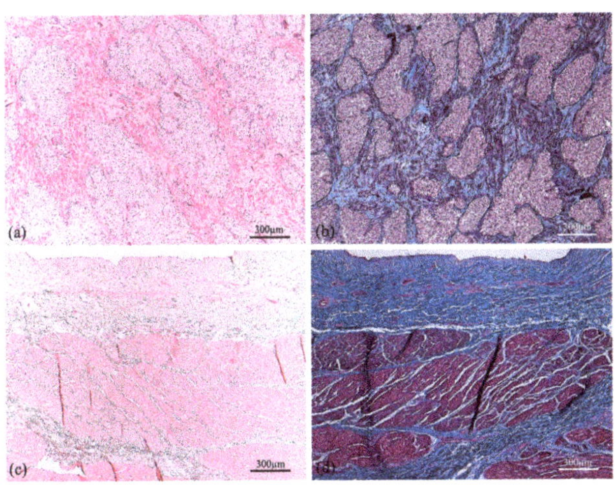

Figure 5. Cross section of human prostate (**a**,**b**) showing tubuloacinar glands surrounded by fibromuscular stroma. Transversal section of the human bladder (**c**,**d**) presenting the mucosa, submucosa, and first muscle layers. Both tissue sections were stained with Haematoxylin–Eosin (**a**,**c**) and Azan–Mallory (**b**,**d**).

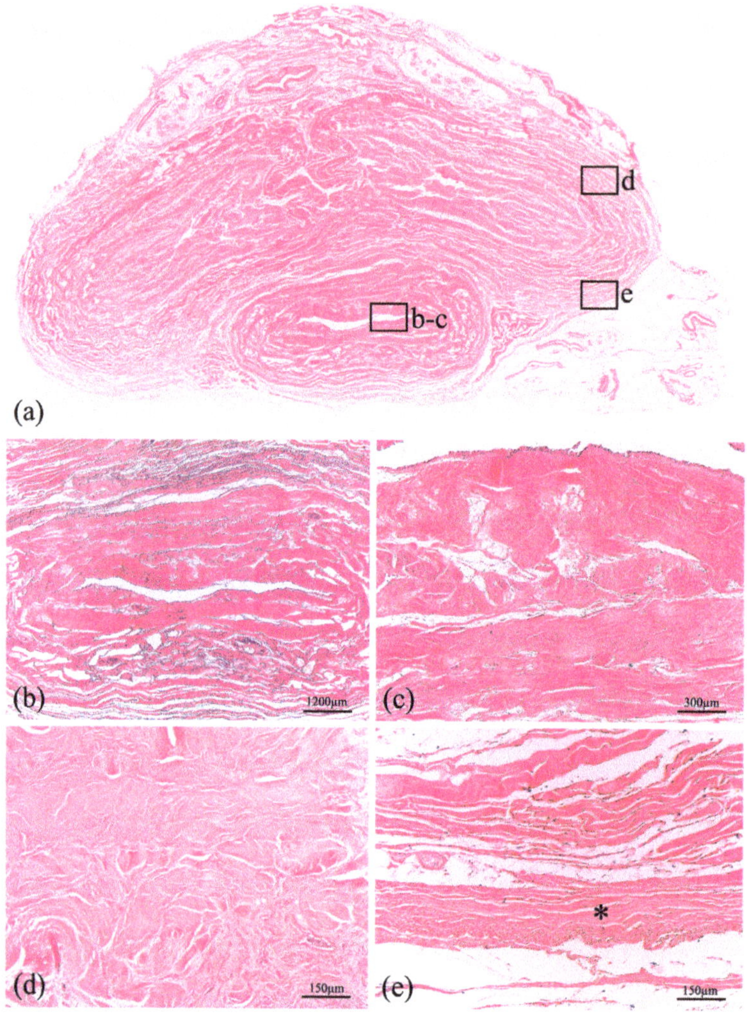

Figure 6. Histological representation of large thickness specimen of the human penis (**a**) stained with Haematoxylin–Eosin. The section at the lower center is the corpus spongiosum surrounding the bulbar urethra, shown at two different magnifications (**b**,**c**). In the upper part of the macrosection, two corpora cavernosa are surrounded by tunica albuginea (**d**). All these structures are covered by Buck's fascia, a deep fascia layer made of fibrous connective tissue, indicated by an asterisk (**e**).

Additional staining (Azan–Mallory) was performed to distinguish muscle fiber content with red color and connective tissue, highlighting collagen fibers with blue color (Figures 5b,d and 7). Therefore, the tunica albuginea (Figure 7d) around the corpora cavernosa can be noticed, as well as the Buck's fascia (Figure 7e), which results in a deep fascia layer made of fibrous connective tissue, indicated by an asterisk. Moreover, the transversal section of the human bladder (Figure 5d) reveals the mucosa, submucosa and first muscle layers.

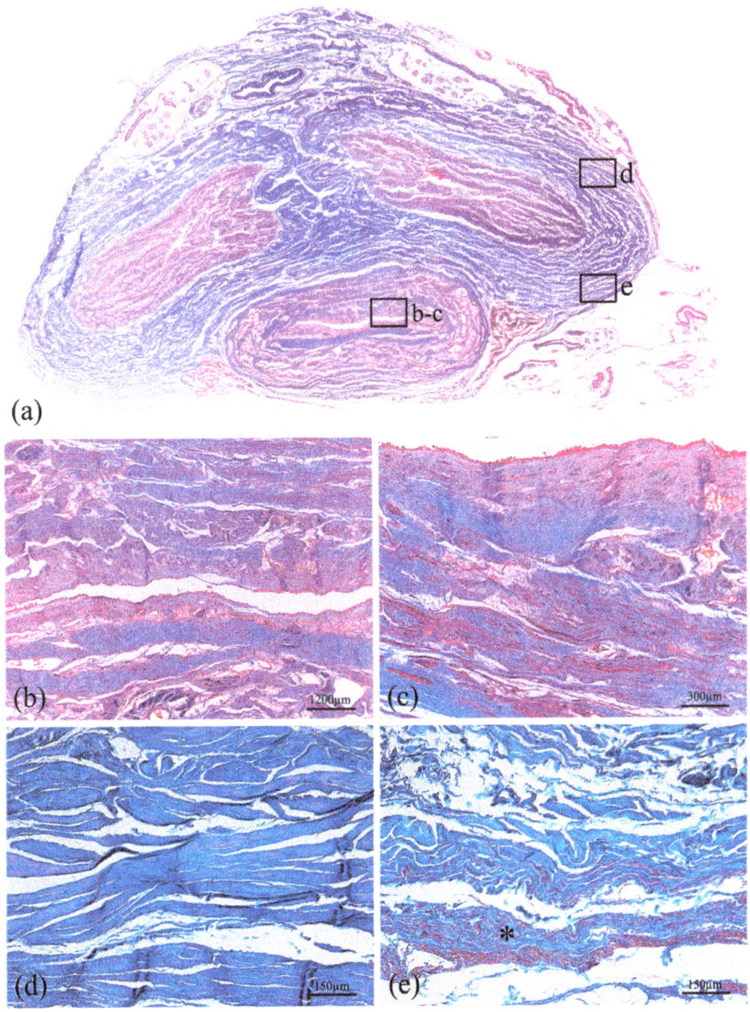

Figure 7. Histological representation of a transversal section of the human penis (**a**) stained with Azan–Mallory. The section at the lower center is the bulbar urethra surrounded by corpus spongiosum and shown at two different magnifications (**b**,**c**). In the upper part of the macrosection, two corpora cavernosa are surrounded by tunica albuginea (**d**). All these structures are covered by Buck's fascia, a deep fascia layer made of fibrous connective tissue, indicated by an asterisk (**e**).

4. Discussion

Human LUT biomechanics research is in its infancy, different from research on many human tissues that have already extensively investigated. On the basis of the literature, currently, few works have been published on the mechanical investigation of some of these biological tissues, while others have never been studied before or only partially.

Moreover, biological tissues are usually characterized by a huge inter-subject variability due to a variety of factors such as subject age, gender, BMI and similar. This usually affects the results when the focus is the quantification of the difference between tissues from the same anatomical district. For this reason, the present study reported the mechanical characterization of the tissues of the LUT harvested from the same subject and tested with the same protocols to quantify each tissue's contribution to the overall mechanical response of the district.

Figure 2 reports the median failure curves that were obtained for the bladder, the Buck's fascia, the prostate and the tunica albuginea, with mechanical characteristics listed in Table 4. As a first analysis, the bladder and prostate exhibited similar tensile behavior in terms of UTS and E of the linear region, while strain at break spans for all the tissues from about 21% up to 85%. Tunica albuginea resulted in the stiffest tissue, with a UTS of about four times greater than the others, with E increasing by one order of magnitude. The bladder showed a restrained variability, suggesting no predominant fibers distribution, while clear differences were observed for the prostate and the tunica, even if, for the latter, the number of samples was limited.

Animal bladder tissues were analyzed by Zanetti et al. [9], who studied the effect of strain rate, orientation and loading history on the behavior in porcine, and then added uniaxial cyclic tests at different strain rates (10, 50, 100, 200, 500% s^{-1}) reaching a strain of about 50% [10]. Contrary to what has been observed in this study, porcine bladders showed greater stiffness along the transverse loading direction and were more influenced by the strain rate than those tested along the apex-to-base ones, even if they observed a homogeneous distribution of collagen fibers in the animal tissue. In the present work, the human bladder strain reached in the loading phase was lower (51.42%, [47.23, 54.69]%) than the strain values reported for porcine, whose loading phase covered a final strain of 100% adopting similar strain rates (1.3% s^{-1} in [9] while 2.5% s^{-1} in this protocol), but with higher recorded stresses (0.47 MPa, [0.44, 0.54] MPa vs about 0.05 MPa [9]). On the basis of the thickness measurements, the pig bladder samples resulted as thicker than the human ones in the present work (3.99 $\pm$ 1.04 mm [9] and 4.37 $\pm$ 1.00 mm [10] vs. 2.14 $\pm$ 0.31 mm).

Our results in human samples are also supported by the histological analysis as well as motivated by their physiological function. In particular, histological staining with Haematoxylin–Eosin (Figure 6) and Azan–Mallory (Figure 7) was performed to identify the gross appearance and more specific tissue composition of human specimens. Figure 7 shows the muscle (red) and connective tissue (blue) distribution, distinguishing between compact tissue near the urethral lumen and porous tissue on the outer side, represented by corpus spongiosum. Concerning urethral tissue, the second magnification in Figure 7b shows the different layers of the urethra transversal section, indicating the main layers and fiber directions. The two different muscle fiber directions confirm the mechanical behavior of the tissue, supporting also the data of Cunnane et al., 2021 [5] and Masri et al., 2018 [4].

To investigate the prostatic fibrosis as responsible for LUT dysfunctions, Ma et al. [15] collected an entire prostate from a cadaver (66 years) and periurethral tissue from 28 men undergoing radical prostatectomy. Similar to the here reported protocol, the prostate was opened along the urethra and slices were taken parallel and perpendicular to the urethra (thickness 3.3 $\pm$ 0.4 mm), while prostatectomy tissues were trimmed in strips, whose thickness (2.0 $\pm$ 0.6 mm) resulted very similar to the measures of the present study (1.96 $\pm$ 0.62 mm). When considering the healthy tissue, they found an average tangent modulus of 450 $\pm$ 77 kPa for the samples parallel to the urethra while 560 $\pm$ 180 kPa for the perpendicular samples, reporting a minor difference between the two directions if compared to the results of the present study, even if adopting similar strain rate (1% s^{-1} up to a maximum strain of 30%), as well as lower absolute values. In the meantime, the prostatectomy tissues reported a wider range of tangent modulus between 9 and 2390 kPa with greater stiffness directly correlated with higher collagen content due to fibrosis and with moderate/severe LUT symptoms. These results could support the evidence of a reparative adaptation of the prostate tissue in response to the endoscopic intervention documented in the donor's health history (TURP), which was characterized by a stiffer response along both directions (Table 4), but maintaining the anisotropic behavior due to a non-homogeneous fibers distribution (Figure 5).

The human tunica albuginea was only partially studied in healthy conditions with unique methods [16], but because of the particular testing technique, it was difficult to compare the results to those obtained in the present study. However, the thickness values were very similar (2.42 ± 0.53 mm [16] vs. 2.38 ± 0.52 mm). Regarding the mechanical basis of pathological conditions, Brady et al. [17] proposed uniaxial tensile tests up to failure (strain rate 1 mm s^{-1}) of the tunica albuginea affected by PD. Specimens exhibited high variability in mechanical properties (E of 5–118 MPa, UTS 1.1–6.1 MPa and strain at UTS 9–42%). These findings showed an altered tissue behavior with mineralization, supported by the literature studies on atherosclerotic plaque mechanics which found higher modulus [19–21] in calcified plaques compared to healthy tissue. Healthy tunica albuginea samples tested in the present study showed mechanical properties close to the lower limit of the wide range shown in [17]. Tensile strength and Young's modulus of tunica albuginea were also evaluated in a primate model by Kandabarow et al. [18], reporting, respectively, an average value of 1.9 ± 0.4 MPa and 11.9 ± 4.2 MPa along the circumferential direction, very close to the values obtained longitudinally in the present study (UTS 2.12 MPa, [1.39, 3.81] MPa; E 12.77 MPa, [10.36, 17.70] MPa).

Referring to the viscoelastic behavior of the human LUT tissues, stress–relaxation tests were performed in this work. Figure 3a–d report the normalized relaxation curves of the urethra when considering different regions (membranous, penile or bulbar), locations (distal or proximal) and directions (longitudinal or transverse). No significant differences were observed varying the regions or the location, with similar percentages of relaxation between about 78% and 88%. The only exception can be seen when comparing the loading direction within the penile region, where a difference of about 10% of relaxation between the longitudinal and the transverse directions was observed. Narrowed confidence intervals showed a good repeatability of the relaxation phenomenon among the samples. Figure 3e showed the comparison with other tissues, i.e., the bladder and the tunica albuginea (only transversal direction was reported due to sample availability). The tunica albuginea had the lowest percentage of relaxation (65.74%, [53.47, 82.12]%), while bladder behavior appeared similar to those of the urethral tissues (88.55%, [84.00, 91.18]%).

Clear anisotropy can be observed between longitudinal and transversal urethral samples when referring to the stress–strain equilibrium curves (Figure 3f). The human urethra exhibited a stiffer response along its axis, while circumferential deformations can be achieved with lower stresses. These insights are also supported by the functional evidence during micturition, where the urine flow applies a radial pressure thus resulting in a predominant radial urethral deformation rather than a longitudinal one. Table 5 reports all the fitting parameters of the different tissues, underlining that, for the equilibrium, the initial stiffness k for the transversal urethral direction is two orders of magnitude lower than the longitudinal one. As during failure, in the equilibrium response, tunica albuginea was the tissue characterized by the greatest initial stiffness (8.64 MPa).

Few other works reported the viscoelastic characterization of human LUT tissues. Masri et al. [4] found that the spongy urethra exhibited more important hysteresis and anisotropy, with a stiffer behavior in the circumferential direction. Contrary to our findings, Cunnane et al. [5] reported no differences between regional responses for circumferential or longitudinal extension and also in terms of the relaxation or creep parameters defining the response in the two directions, while Natali et al. [8] reported greater stiffness for the proximal specimens of equine urethra along the longitudinal direction, as in the present study. A significant difference in terms of stiffness resulted when comparing the longitudinal proximal specimens to all the other samples. Comparing thickness measurements of human urethral strips, the mean value recorded in the present study (3.73 ± 0.78 mm) is within the range outlined by the literature studies mentioned above (2.53 ± 0.24 mm [4], 4.44 ± 0.64 mm [5]).

Stress–relaxation results on porcine bladder samples reported by Zanetti et al. [9] highlighted a longer relaxation time for animal samples (>600 s) with respect to human samples (after 300 s the relaxation phenomenon lasted). In agreement with the present study, they found no differences among the loading directions in terms of viscosity. Even in humans, the regular spatial distribution of the contractile and support elements has been observed in the various layers of the bladder wall (Figure 5). The bladder can be seen as a three-dimensional multiplanar interweaving structure that allows the forces to be uniformly distributed both in expansion and contraction. This conformation ensures a gradual and coordinated modification of bladder diameters during physiological activity, with greater involvement of the dome and lateral walls.

Finally, to investigate LUT biomechanics under additional loading combinations, the compression behavior of the prostate was also analyzed through indentation tests. Figure 4 shows the recorded force between 0 and 1.5 mm of indentation depth and the curves obtained by fitting the experimental data with the elastic Hertz model (Table 6, $R^2 > 0.85$ for all the points with only P3 as an exception). Since the prostate reported some signs of TURP with portions of the gland tissue covered by nodules, points P1, P2 and P7 investigated the stiffness of the nodule surface, then compared with the other points. On average, nodule elastic modulus resulted in 46 $\pm$ 27 kPa, three times greater with respect to the other regions without repair signs (16 $\pm$ 5 kPa). This is consistent with what is known about the repair processes of damaged tissues by surgical procedures, such as endoscopic resection for the clinical diagnosis of benign prostatic hyperplasia in the present case. As expected, the proliferation of connective tissue with a share of elastic fibers consequently leads to different behavior of the repaired tissue compared to the native one. Other studies in the literature reported indentation on prostate glands, especially in pathological conditions. In particular, our outcomes resulted in good agreement with Ahn et al. [13], who performed indentation tests on prostate specimens from 46 patients undergoing radical prostatectomy by adopting a minimally motorized indenter with a hemisphere tip and cylindrical probe (diameter 2 mm) and a similar protocol (3 mm depth indentation at a speed of 1 mm s^{-1}). The Hertz–Sneddon equation was also applied to estimate the elastic modulus with Poisson's ratio of 0.499. The results showed that the region containing cancerous tissue increased the elastic modulus (24.1 $\pm$ 14.5 kPa vs. 17.0 $\pm$ 9.0 kPa). In addition, Kim et al. [14] investigated the relationship between prostate elasticity and LUT dysfunctions. According to indentation results on 48 prostates from patients (age 65.8 $\pm$ 5.9 years) who underwent radical prostatectomy, the median elastic modulus was 20.8 kPa [5.6–22.9] kPa), and the incidences of the LUT dysfunctions were significantly higher in patients with an elastic modulus > 20 kPa.

As this work is the first comprehensive analysis of LUT biomechanics able to exclude possible bias due to inter-subject variability, there are some limitations that should be mentioned. First of all, the number of samples obtained for each tissue in some cases was scarce and thus no statistical analysis was provided as additional support to these findings. However, even if one single case study could result in limited sample availability, narrowed confidence intervals could confirm the advantage of comparing tissues from the same source. Indeed, within this study, attention was primarily paid to underlining the difference between tissues mechanical behavior, functions and composition, and it is possible to conclude that the results clearly highlighted differences and similarities (Figures 2 and 3).

Differences among the experimental setup could have affected some comparisons. Rings vs rectangular samples for tensile tests on the urethra were prepared differently (i.e., for the second group, a sample holder was provided and the samples were attached by means of strong rapid glue). However, no specific sample damage or alteration was observed at the end of the test for either setup, suggesting suitable results for both procedures.

Sample storage could be seen as another limitation. However, after sample harvesting from the cadaver (−80 °C), all the patches were kept at −20 °C, a temperature that was reported to not significantly influence LUT tissues' mechanical performances [7]. Moreover, all the tests were performed in a limited interval of time; thus, samples were stored at this temperature for no more than two months. Comparison with the literature confirmed the reliability of the obtained results.

5. Conclusions

The contribution of the work here proposed is two-fold and it is suitable to fill a clearly missing piece in the literature: for the first time, mechanical tests were carried out on human Buck's fascia, and all the LUT tissues for the mechanical investigation were harvested from the same subject with the potential to highlight the characteristic behavior of the different tissues avoiding the variability due to tissues from different patients.

Currently, this case study sets the basis for the investigation approach for future developments which would require the extension of the numerosity of the population samples to improve the experimental data reliability. Characterization of the essential mechanical properties of the LUT may represent a key aspect for the development of effective in silico models to extend the experimental results to high-level scenarios providing additional information on LUT dysfunctions and reliable diagnostic and/or therapeutic tools for the urological clinical and surgical context [26–28]. The study and validation process of innovative procedures and devices for the treatment of LUT pathologies, such as balloon catheters for dilating urethral strictures or artificial sphincters for restoring urinary continence condition [29–33], needs to be considered according to the mechanical behavior of the biological tissues and structures involved, as well as the development and suitability of phantoms mimicking the tissue performance for surgical planning and training [34] and the tissue engineering for surgical reconstruction of anatomical or functional defects [35,36].

Author Contributions: Conceptualization, A.B., R.B.-B. and E.L.C.; methodology, A.B., M.V.M., C.G.F. and R.B.-B.; software, A.B. and M.V.M.; validation, A.B., M.V.M., C.G.F. and R.B.-B.; formal analysis, M.V.M.; investigation, A.B., M.V.M., M.C. and M.T.; resources, A.P., V.M., R.D.C. and E.L.C.; data curation, A.B., M.V.M. and R.B.-B.; writing—original draft preparation, A.B. and M.V.M.; writing—review and editing, A.B., M.V.M., C.G.F., M.C., M.T., R.B.-B. and E.L.C.; supervision, A.B., C.G.F., R.B.-B. and E.L.C.; project administration, A.P., V.M. and E.L.C.; funding acquisition, C.G.F. and E.L.C. All authors have read and agreed to the published version of the manuscript.

Funding: This work was supported by MIUR, FISR 2019, Project no. FISR2019_03221, titled CECOMES: CEntro di studi sperimentali e COmputazionali per la ModElliStica applicata alla chirurgia and by University of Padova, Project FONT_BIRD2020_01, titled Characterization of Artificial Urinary Sphincters for the identification of new DEVices (AUS-DEV).

Institutional Review Board Statement: All procedures were conducted on human bodies from the "Donation to Science" Body Donation Program of the University of Padova and Veneto Region/National Reference Center for preserving and using gifted bodies. The procedures were carried out in accordance with national laws, ethical standards of the regional/national research committees, and the 1964 Helsinki Declaration, including its later amendments or comparable ethical standards. Participants provided written informed consent to join the Body Donation Program, which has met the requirements of standard ISO 9001:2015 (Registration number: IT-62435-16764) (ISO, 2008). The privacy rights of human subjects were and will always be respected.

Informed Consent Statement: Informed consent was obtained from all subjects involved in the study, consistently with the setting of collection/donation of the anatomical material.

Data Availability Statement: The raw/processed data are available, contacting the corresponding authors, on reasonable request. The article contains all the necessary health data of the body donors, which can/should be utilized for scientific research purposes based on the informed consent provided. Any additional information that is not relevant or useful to the research is kept private and not disclosed to protect the privacy of the individuals.

Conflicts of Interest: The authors declare no conflicts of interest.

References

1. Mahadevan, V. Anatomy of the lower urinary tract. *Surgery* **2016**, *34*, 318–325. [CrossRef]
2. Dunnick, N.R.; Sandler, C.M.; Jeffrey, H. *Newhouse, Textbook of Uroradiology*, 5th ed.; Lippincott Williams & Wilkins: Philadelphia, PA, USA, 2012.
3. Singh, G.; Chanda, A. Mechanical properties of whole-body soft human tissues: A review. *Biomed. Mater.* **2021**, *16*, 062004. [CrossRef] [PubMed]

4. Masri, C.; Chagnon, G.; Favier, D.; Sartelet, H.; Girard, E. Experimental characterization and constitutive modeling of the biomechanical behavior of male human urethral tissues validated by histological observations. *Biomech. Model. Mechanobiol.* **2018**, *17*, 939–950. [CrossRef] [PubMed]
5. Cunnane, E.M.; Davis, N.F.; Cunnane, C.V.; Lorentz, K.L.; Ryan, A.J.; Hess, J.; Weinbaum, J.S.; Walsh, M.T.; O'Brien, F.J.; Vorp, D.A. Mechanical, compositional and morphological characterisation of the human male urethra for the development of a biomimetic tissue engineered urethral scaffold. *Biomaterials* **2021**, *269*, 120651. [CrossRef] [PubMed]
6. Lalla, M.; Danielsen, C.C.; Austevoll, H.; Olsen, L.H.; Jørgensen, T.M. Biomechanical and Biochemical Assessment of Properties of the Anterior Urethra After Hypospadias Repair in a Rabbit Model. *J. Urol.* **2007**, *177*, 2375–2380. [CrossRef] [PubMed]
7. Cunnane, C.V.; Croghan, S.M.; Walsh, M.T.; Cunnane, E.M.; Davis, N.F.; Flood, H.D.; Mulvihill, J.J. Cryopreservation of porcine urethral tissue: Storage at −20°C preserves the mechanical, failure and geometrical properties. *J. Mech. Behav. Biomed. Mater.* **2021**, *119*, 104516. [CrossRef]
8. Natali, A.N.; Carniel, E.L.; Frigo, A.; Pavan, P.G.; Todros, S.; Pachera, P.; Fontanella, C.G.; Rubini, A.; Cavicchioli, L.; Avital, Y.; et al. Experimental investigation of the biomechanics of urethral tissues and structures. *Exp. Physiol.* **2016**, *101*, 641–656. [CrossRef]
9. Zanetti, E.M.; Perrini, M.; Bignardi, C.; Audenino, A.L. Bladder tissue passive response to monotonic and cyclic loading. *Biorheology* **2012**, *49*, 49–63. [CrossRef]
10. Natali, A.N.; Audenino, A.L.; Artibani, W.; Fontanella, C.G.; Carniel, E.L.; Zanetti, E.M. Bladder tissue biomechanical behavior: Experimental tests and constitutive formulation. *J. Biomech.* **2015**, *48*, 3088–3096. [CrossRef]
11. Krouskop, T.A.; Wheeler, T.M.; Kallel, F.; Garra, B.S.; Hall, T. Elastic Moduli of Breast and Prostate Tissues Under Compression. *Ultrason. Imaging* **1998**, *20*, 260–274. [CrossRef]
12. Phipps, S.; Yang, T.H.J.; Habib, F.K.; Reuben, R.L.; McNeill, S.A. Measurement of the mechanical characteristics of benign prostatic tissue: A Novel method for assessing benign prostatic disease. *Urology* **2005**, *65*, 1024–1028. [CrossRef] [PubMed]
13. Ahn, B.M.; Kim, J.; Ian, L.; Rha, K.H.; Kim, H.J. Mechanical property characterization of prostate cancer using a minimally motorized indenter in an ex vivo indentation experiment. *Urology* **2010**, *76*, 1007–1011. [CrossRef]
14. Kim, K.H.; Ahn, B.; Lim, S.K.; Han, W.K.; Kim, J.H.; Rha, K.H.; Kim, J. Indenter study: Associations between prostate elasticity and lower urinary tract symptoms. *Urology* **2014**, *83*, 544–549. [CrossRef]
15. Ma, J.; Gharaee-Kermani, M.; Kunju, L.; Hollingsworth, J.M.; Adler, J.; Arruda, E.M.; Macoska, J.A. Prostatic fibrosis is associated with lower urinary tract symptoms. *J. Urol.* **2012**, *188*, 1375–1381. [CrossRef]
16. Bitsch, M.; Kromann-Andersen, B.; Schou, J.; Sjøntoft, E. The Elasticity and the Tensile Strength of Tunica Albuginea of the Corpora Cavernosa. *J. Urol.* **1990**, *143*, 642–645. [CrossRef] [PubMed]
17. Brady, L.; Stender, C.J.; Wang, Y.-N.; Schade, G.R.; Maxwell, A.D.; Wessells, H.; Ledoux, W.R. Mechanical characterization of fibrotic and mineralized tissue in Peyronie's disease. *Int. J. Impot. Res.* **2022**, *34*, 477–486. [CrossRef]
18. Kandabarow, A.M.; Chuang, E.; McKenna, K.; Le, B.; McVary, K.; Colombo, A. Tensile strength of penile tunica albuginea in a primate model. *J. Urol.* **2021**, *206*, 1073. [CrossRef]
19. Sherebrin, M.H.; Bernans, H.A.; Roach, M.R. Extensibility Changes of Calcified Soft Tissue Strips from Human Aorta. *Can. J. Physiol. Pharmacol.* **1987**, *65*, 1878–1883. [CrossRef]
20. Walraevens, J.; Willaert, B.; De Win, G.; Ranftl, A.; De Schutter, J.; Sloten, J.V. Correlation between compression, tensile and tearing tests on healthy and calcified aortic tissues. *Med. Eng. Phys.* **2008**, *30*, 1098–1104. [CrossRef]
21. Loree, H.M.; Grodzinsky, A.J.; Park, S.Y.; Gibson, L.J.; Lee, R.T. Static circumferential tangential modulus of human athereosclerotic tissue. *J. Biomech.* **1994**, *27*, 195–204. [CrossRef]
22. Porzionato, A.; Macchi, V.; Stecco, C.; Mazzi, A.; Rambaldo, A.; Sarasin, G.; Parenti, A.; Scipioni, A.; De Caro, R. Quality management of Body Donation Program at the University of Padova. *Anat. Sci. Educ.* **2012**, *5*, 264–272. [CrossRef]
23. Biomomentum Catalog. Available online: https://www.biomomentum.com/catalog/ (accessed on 1 September 2023).
24. Bonaldi, L.; Berardo, A.; Pirri, C.; Stecco, C.; Carniel, E.L.; Fontanella, C.G. Mechanical Characterization of Human Fascia Lata: Uniaxial Tensile Tests from Fresh-Frozen Cadaver Samples and Constitutive Modelling. *Bioengineering* **2023**, *10*, 226. [CrossRef]
25. Zwirner, J.; Babian, C.; Ondruschka, B.; Schleifenbaum, S.; Scholze, M.; Waddell, N.J.; Hammer, N. Tensile properties of the human iliotibial tract depend on height and weight. *Med. Eng. Phys.* **2019**, *69*, 85–91. [CrossRef]
26. Mascolini, M.V.; Fontanella, C.G.; Berardo, A.; Carniel, E.L. Influence of transurethral catheters on urine pressure-flow relationships in males: A computational fluid-dynamics study. *Comput. Methods Programs Biomed.* **2023**, *238*, 107594. [CrossRef]
27. Herwig, R.; Bayerl, M. Superficial tunica albuginea rupture as initial starting point of Peyronie's disease: A topic for interdisciplinary consideration. *Biomed. Res. Int.* **2015**, *2015*, 751372. [CrossRef] [PubMed]
28. Wang, F.; Cao, Z.; Zhai, L.; Zhang, J.; Kong, H.; Lin, W.; Fan, Y. Biomechanical study of the male lower urinary tract: Simulation of internal and external sphincters dyssynergia. *J. Biomech.* **2023**, *149*, 111475. [CrossRef] [PubMed]
29. Toniolo, I.; Mascolini, M.V.; Carniel, E.L.; Fontanella, C.G. Artificial sphincters: An overview from existing devices to novel technologies. *Artif. Organs* **2022**, *47*, 617–639. [CrossRef] [PubMed]
30. Natali, A.N.; Fontanella, C.G.; Carniel, E.L. Biomechanical analysis of the interaction phenomena between artificial urinary sphincter and urethral duct. *Int. J. Numer. Method. Biomed. Eng.* **2020**, *36*, e3308. [CrossRef] [PubMed]

31. Mascolini, M.V.; Fogang, J.V.F.; Salomoni, V.; Fontanella, C.G.; Carniel, E.L.; Natali, A.N. A novel artificial urinary sphincter and comparison with reference device. In Proceedings of the 8th National Congress of Bioengineering, GNB 2023, Padova, Italy, 21–23 June 2023; Patron Editore S.r.l.: Bologna, Italy, 2023.
32. Bhave, A.; Sittkus, B.; Urban, G.; Mescheder, U.; Möller, K. Finite element analysis of the interaction between high-compliant balloon catheters and non-cylindrical vessel structures: Towards tactile sensing balloon catheters. *Biomech. Model. Mechanobiol.* **2023**, *22*, 2033–2061. [CrossRef] [PubMed]
33. De Menech, Q.; Konstantinidi, S.; Martinez, T.; Benouhiba, A.; Civet, Y.; Perriard, Y. Mechanical Characterisation of Porcine Urethra: Non Linear Constitutive Models and Experimental Approach. In Proceedings of the International Conference on Advances in Biomedical Engineering, Dubai, UAE, 16–17 December 2023; ICABME, Institute of Electrical and Electronics Engineers Inc.: NewYork, NY, USA, 2023; pp. 35–40. [CrossRef]
34. Verstraeten, M.; Kheir, G.B.; Perre, L.V.; Raffoul, R.; Cerda, J.C.; Delchambre, A.; Roumeguere, T.; Vanhoestenberghe, A.; Nonclercq, A. Urinary bladder phantom mimicking mechanical properties and pressure during filling. *Biomed. Phys. Eng. Express* **2023**, *9*, 055006. [CrossRef] [PubMed]
35. Sartoneva, R.; Lyyra, I.; Juusela, M.; Sharma, V.; Huhtala, H.; Massera, J.; Kellomäki, M.; Miettinen, S. In vitro biocompatibility of polylactide and polybutylene succinate blends for urethral tissue engineering. *J. Biomed. Mater. Res. B Appl. Biomater.* **2023**, *111*, 1728–1740. [CrossRef] [PubMed]
36. Cheng, Q.; Zhang, L.; Zhang, J.; Zhou, X.; Wu, B.; Wang, D.; Wei, T.; Shafiq, M.; Li, S.; Zhi, D.; et al. Decellularized Scaffolds with Double-Layer Aligned Microchannels Induce the Oriented Growth of Bladder Smooth Muscle Cells: Toward Urethral and Ureteral Reconstruction. *Adv. Healthc. Mater.* **2023**, *12*, e2300544. [CrossRef] [PubMed]

applied sciences

MDPI

Review

Mechanical Stapling Devices for Soft Tissue Repair: A Review of Commercially Available Linear, Linear Cutting, and Circular Staplers

Vito Burgio [1], Janira Bei [1], Mariana Rodriguez Reinoso [1], Marco Civera [1], Oliver Grimaldo Ruiz [1], Cecilia Surace [1] and Nicola M. Pugno [2,3,*]

1 Laboratory of Bio-Inspired Nanomechanics, Department of Structural, Geotechnical and Building Engineering, Politecnico di Torino, 10129 Turin, Italy; vito.burgio@polito.it (V.B.); janirabei7@gmail.com (J.B.); mariana.rodriguez@polito.it (M.R.R.); marco.civera@polito.it (M.C.); oliver.ruiz0809@gmail.com (O.G.R.)

2 Laboratory for Bioinspired, Bionic, Nano, Meta, Materials & Mechanics, Department of Civil, Environmental and Mechanical Engineering, University of Trento, Via Mesiano 77, 38123 Trento, Italy

3 School of Engineering and Materials Science, Queen Mary University of London, Mile End Road, London E1 4NS, UK

* Correspondence: nicola.pugno@unitn.it

Abstract: Stapling devices have emerged as a widespread and effective option for soft tissue surgery, offering promising outcomes for patients by reducing complication rates and surgery time. This review aims to provide an exhaustive analysis of commercially available alternatives in the market, incorporating insights from market analysis, patent landscape, and the existing literature. The main focus lies in identifying and evaluating the most widely adopted and innovative stapling devices, including linear, linear cutting, circular, and powered staplers. In addition, this review delves into the realm of bioabsorbable staples, exploring the materials utilized and the surgical fields where these advanced staples find applications. To facilitate easy comprehension, the gathered information is presented in tables, highlighting the essential parameters for each stapling device. This comprehensive research about stapling devices is intended to aid healthcare practitioners and researchers in making informed decisions when choosing the most appropriate instrument for specific surgical procedures.

Keywords: staplers; surgical stapler; linear; circular; anastomosis; surgical instruments

1. Introduction

In this work, surgical staplers, which are internal use devices employed to deliver staplers to tissues inside the body during surgery for removing part of an organ, cutting through and sealing organs, and creating connections between structures, are covered.

Modern surgical staplers are designed to be disposable, with a maximum number of firing actions, as well as cartridges and other products used for surgical stapling. Today's staples have a rectangular cross-section and become "B" shaped when compressed against the anvil of the stapler to reduce the risk of ischemia, allowing blood to flow through them to the tissue edges. The specific geometry has been developed to ensure hemostasis and to allow sufficient vascularization of the tissue, supporting wound healing [1].

Three major types of stapling devices used both in open and laparoscopic surgery can be identified: linear, linear cutting, and circular. Linear instruments form a straight staple line and may or may not include a blade, while circular instruments have staples set in a crown shape and always include a blade. The ever-increasing diffusion of laparoscopy and robotic surgery led to the diffusion of stapling devices dedicated to endoscopic techniques; linear and circular stapling devices are employed mostly in gastrointestinal procedures such as complex gasterectomies, bariatric treatments, and colorectal oncological resections.

Devices have been compared in terms of procedure duration, hospital stay, wound infection, anastomotic leakage, and anastomotic structures.

The outcome of the repair made by staples or suture techniques has attracted much interest in the scientific community. Recently, the mechanical behavior of repairs made with staples, sutures, and hybrid methods (staple and suture together) was evaluated carefully, showing that staples can resist high loads but are less deformable and rigid than handsewn sutures. This suggests safer employment in the case of small defects or diastasis, where the presumed tissue displacement is minimal. A hybrid repair improves its efficiency, becoming crucial in case of larger defects where the expected tissue displacement is maximal [2].

Although being more complex to operate and resulting in longer procedure times, in several studies covering gastric, bariatric, and colorectal surgery, linear staplers appeared to be preferred in clinical use due to their lower rate of anastomosis leakage and anastomosis strictures, with a sensible reduction in patients' postoperative morbidity [3–6].

Regardless of its specific configuration, a surgical stapler is always characterized by an actuation handle: the surgeon can operate the device manually through a sequence of squeezing a handle or sliding a knob. The effective function of stapling, cutting, and ejecting the staples is performed by the loading unit: two jaws are coupled in different ways to act, one as a cartridge loader for the staples and the other as an anvil to allow the closing of the staples. In the case of circular staplers, the anvil jaw is detachable, granting its insertion into the anastomosis site. The actuation handle and the loading unit are usually connected to each other: the input from the surgeon is translated into the mechanical or electro-mechanical firing system through the use of various keyed shafts [7]. Additional pins can be inserted into the jaws or be comprised into the loading unit to serve as alignment or safety devices, ensuring the correct compression of the tissue before firing or avoiding retention after staple delivery.

Each device is completely in accordance with the aims of surgical stapling, which consist, from a medical point of view, of creating an adequate lumen, preserving adequate vascularization, preventing tension in tissues, avoiding leakage, providing hemostasis, and ensuring mechanical reliability [8].

Non-absorbable staples are the ones most widely used in clinics thanks to their low artifact production in CT and their non-magnetic behavior, which causes minimal distortions in MRI. Titanium is the leading material, both pure or in alloys, thanks to its high resistance and lighter weight compared to stainless steel. Furthermore, titanium is corrosion-resistant but, most importantly, anti-allergic, and it can also be employed in patients suffering from chromate-nickel allergies [1,9].

The general design of stapling devices includes an anvil, a cartridge assembly to support an array of staples, a mechanism to push the cartridge close to the anvil, an alignment pin for capturing tissue between the cartridge and the anvil, and a firing mechanism for ejecting the surgical staples. The firing process is guided by a sequence of pressing the handle: the first press grants tissue compression and allows the surgeon to verify the correct alignment of the jaws, while the second press releases the staples. A button placed on the stapler permits the opening of the jaws at the end of the operation, setting the tissue free and allowing the reloading process.

The main particularity of cutting staplers is the addition of a blading element that allows for making a resection in the middle of the rows of staples. Instead, for endo cutting staplers, the instrument is composed of a disposable articulating end effector comprising the cartridge, the anvil, and a disposable handling unit containing the firing mechanism and the safety control system. The circular staplers have the same components with the particularity of a circular anvil and arrangement of the staples with a blading element. The powered stapling devices have the distinction of a powered actuation, meaning that the firing mechanism is controlled by an electro-mechanical system that gains energy from a battery. Figure 1 shows a description and representation of each type of device.

	General description	Schematic representation
Linear Stapler	This device is a surgical device with control buttons on the handle to prevent accidental firing. It typically features a retention pin or lever for manual or automatic actuation. The device includes buttons for aligning the cartridge, positioning tissue, and opening jaws for tissue release and reloading after the operation.	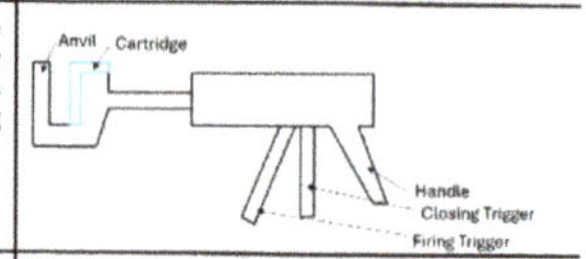
Linear Cutting Stapler	This device features a sliding knob for activation, aligning and firing the blade simultaneously with stapling. The cutting line is intentionally shorter for safety. A release button disengages the cartridge after firing, and a safety tag covers the blade to prevent accidental firing during reloading.	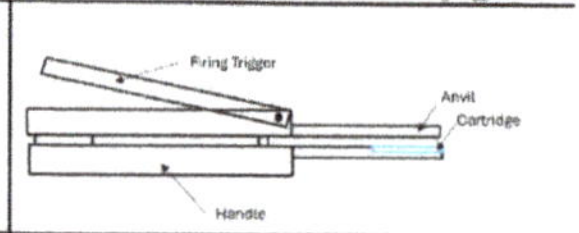
Linear Endo Stapler	This device has an articulating end effector attached through a rotating locking system. After correct alignment, the handle is pushed to close jaws over tissue. A side button unlocks the firing mechanism, and continuous handle pressure delivers staples while the blade moves forward, gradually compressing and cutting the tissue.	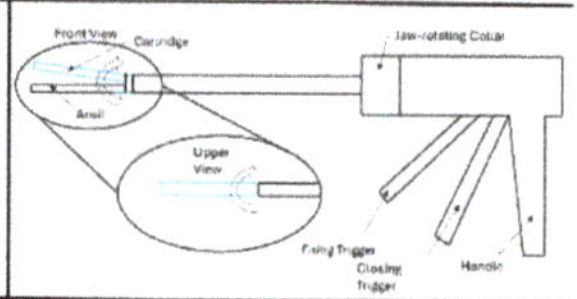
Circular Stapler	This device consists of a detachable anvil and a handling part with a trocar for the cartridge, cutting element, rotary knob for jaw control, and an operating handle. The process involves opening the rotary knob to extract the anvil, detaching it for tissue insertion, and then assembling it with the cartridge by rotating the knob until an indicator confirms sufficient tissue compression.	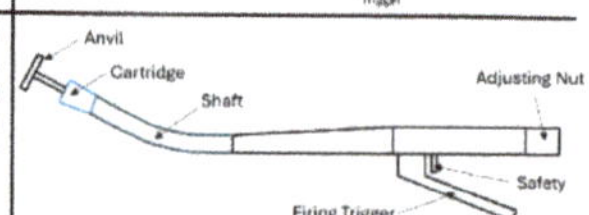
Powered Stapler	This device features battery-driven actuation, eliminating manual control. The firing mechanism is powered by a battery, operating an electro-mechanical system that controls the cutting element and the cartridge..	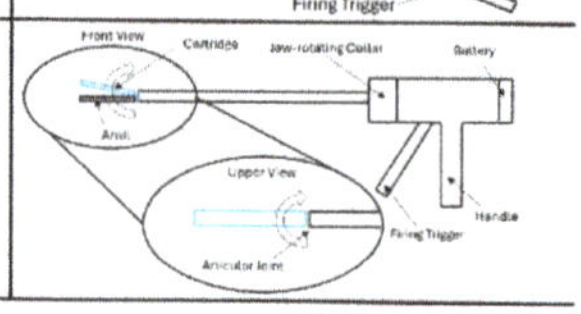

Figure 1. Description of key components for each type of device. The cartridge is highlighted in blue.

Absorbable Staples

The use of absorbable staples is mainly spreading in dermal suturing and in performing ligations [10–15]. Currently, considering the bariatric, gastrointestinal, vascular, and thoracic surgery fields, which are the main fields of application of staplers, the only commercially available absorbable staple is Medtronic Polysorb [16–18], while other bioresorbable magnesium and zinc alloys are still a matter of research. Although studies claim to match the clinical requirements for a secure anastomosis, their technology is yet to be implemented by manufacturers on a production scale [19,20].

Another field of application for staplers is deep dermal suturing. In particular, in plastic surgery, the use of staplers has been adopted and provides wound healing outcomes that are better than those of conventional suturing [15].

Staplers are also used in orthopedics to secure soft tissues to bone [21–23]. Few applications have been found comprising staples in the process of suturing soft tissues in meniscal or rotator cuff repair, the latter involving scaffold placement [24,25].

When comparing devices of each type, a variety of features must be considered since each manufacturer proposes its own combination of specifications. In this work, all available characteristics of surgical staplers will be presented to compare the available alternatives for the three different classes of staplers.

2. Materials and Methods

The first step was the analysis of the current surgical stapling market through ECRI Institute archives, which allowed us to identify the main competitors on the global scene (Ethicon (Johnson & Johnson, New Brunswick, NJ, USA) endo-Surgery and Covidien (Medtronic, Minneapolis, MN, USA)). The Italian market was then assessed using information accessed from Ministry of Health reports. In light of the Italian public health system's data, Panther Medical (Beiqijia, Beijing, China) and Sinolinks (Zhonglou Zone, Changzhou, China) were taken into consideration for the analysis.

After choosing the manufacturers, their websites were consulted, and all the available technical documentation was extracted. The main parameters to make a comparison were identified as staple line length for linear devices, staple line diameter for circular devices, open and closed staple height, the maximum number of firings in case of reloadable devices, the maximum number of rows of staples placed on each side of the suturing line, and length of the shaft in case of Endo-related devices.

Each device was identified both by its commercial name and by its product code, allowing for a better univocal identification. To further explore the design and the functioning mechanism of each product, some of the patents have been assessed through OrbitExpress online (www.questel.com) with the search query shown in Table 1 (the search was done in January 2021). Among the 416 retrieved results, 216 have been included based on their International Patent Classification, these being A61B-017/068, A61B-017/072, or A61B-017/115, classes specifically dedicated to surgical staplers. Patents have been scanned on the basis of title and abstract and on the basis of their latest available and granted version, resulting in 38 included documents. The decisional process is described in Figure 2. For all other devices, it was not possible to identify the patent documentation because of a lack of correspondence between commercial and technical denominations or because of intellectual property restrictions and trade secrets. Finally, specifications coming from the FDA medical devices database, which were assessed using manufacturers' names and product codes as keywords for the query, were recorded.

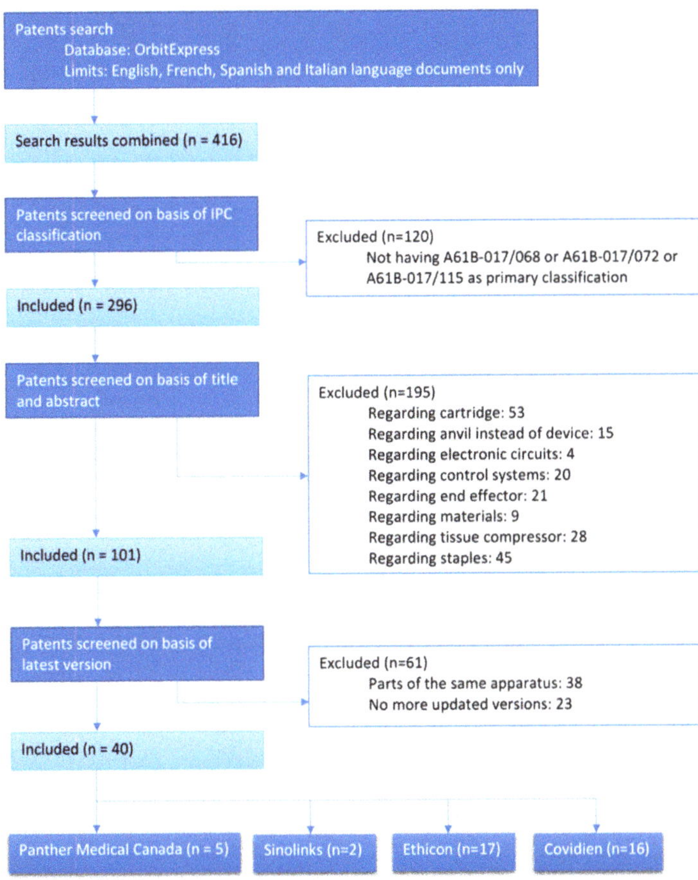

Figure 2. Patent inclusion flowchart.

Table 1. Patent search queries using OrbitExpress Questel software.

Keywords	Surgical stapler or stapler or stapling device or stapling apparatus or endo stapler or circular stapler or linear stapler or cutting stapler
Assignee	Ethicon Endo-Surgery or Covidien or Panther Medical or Sinolinks
Classification	A61B-017/068 or A61B-017/072 or A61B-017/115
Publication Country	WO or US or EP

3. Results

In this section, the information collected about the different classes of staplers was divided into subsections to organize the findings better. The crucial characteristics of the staplers were organized in tables to summarize. When some information was not available on the website an n.a (not available) was reported.

3.1. Linear Staplers

Linear devices are employed mostly in gastrointestinal procedures [26]. Other uses of linear staplers are abdominal surgery, gynecology, vascular surgery, and pediatric surgery.

Table 2 summarizes the principal characteristics of linear staplers in commerce by manufacturer. In particular, the commercial name and product code are reported according to the manufacturer. Other features that are reported include the staple line length, with values ranging between 30 mm and 60 mm for all manufacturers, while 45 mm and 90 mm sizes are proposed by everyone except Ethicon. Staple height depends on the cartridge mounted on the stapler: values vary from 2.5 mm proposed by Ethicon and Covidien for vascular applications to 4.8 mm for all the manufacturers for thicker tissues. For applications with a medium thickness, 3.5 mm is the chosen value, with only Sinolinks and Ethicon offering 3.8 mm and 3.85 mm alternatives. The height of the closed staple is quite standardized: 1.5 mm is ensured by all manufacturers, and 2 mm is available in every catalog except from Sinolinks, which offers a 1.8 mm alternative. Several rows placed by the instruments are also standardized—only Covidien and Ethicon propose a version of the stapler specific for vascular applications, which grants a height of 1mm but requires three rows of staples instead of two. The exceeding row is determined to be useful in providing a secure anastomosis and avoiding blood leakage issues.

Several studies in the literature analyzed the effects of linear stapler suturing and handsewn methods [26–28]. In a review, Choy et al. [27] compare linear stapled versus handsewn methods for ileocolic anastomosis. The authors analyzed different outcome measures like clinical and radiological anastomotic leakage, anastomotic stricture, hemorrhage, and time. Their results showed that stapled functional end-to-end ileocolic anastomosis is associated with fewer anastomotic leaks than handsewn anastomosis and should be considered the gold standard.

A meta-analysis of Gong et al. [26] analyzes stapled vs. hand suture closure of loop ileostomy. The authors affirm that stapled loop ileostomy reversal is superior to hand sutures in reducing postoperative small bowel obstruction, operation time, and hospital stay, and therefore, stapled side-to-side anastomosis should be considered the preferred technique.

According to the study of Giaccaglia et al. [9], in which three linear staplers of different companies were compared, small differences emerged between the devices regarding anastomotic pressure resistance and tensile strength. Many studies have confirmed the advantages of using linear staplers over circular staplers. Regarding a study on 213 patients with gastric cancer who underwent laparoscopic total gastrectomy, although mean operation time in the linear stapler group was longer than the circular stapler group, using linear staplers, anastomosis leakages and costs related to anastomosis were lower, and anastomosis stenosis did not occur. These advantages led scientists to prefer linear staplers in laparoscopic total gastrectomy [6].

Linear staplers are more effective than circular staplers in laparoscopic gastrojejunal (GJ) anastomosis in morbid obesity surgery [3].

Table 2. Linear stapler characteristics.

Brand	Commercial Name	Code	Staple Line Length [mm]	Staple Height [mm]	No. of Rows	Closed Staple Height [mm]	Firing Actions	Surgery Application and Relevant Info	Reference
PANTHER HEALTHCARE	Linear Stapler	FLSLF30 FLSLF45 FLSLF60 FLSLF90	30.0 45.0 60.0 90.0	3.50 4.80	2	1.5 2.0	It can be reloaded up to 7 times for a total of 8 firings in a single procedure.	n.a	(Panther Healthcare, Open Linear Staplers) [29]
SINOLINKS	Disposable Linear Stapler	DLS B DLS C DALS A DLS A	30.0	3.80 4.80	n.a	1.5 1.8	n.a	In gastrointestinal surgery, they are used for closure of the stump or incision in digestive tract reconstruction and viscera resection. Features: - DLS C: Three rows of staples provide better hemostasis. - DALS A: Two-stage closure for easy tissue adjustment. - DLS A: Manual integrated tissue positioning needle and flexible operation.	(Sinolinks, Linear Staplers)
SINOLINKS	Disposable Linear Stapler	DLS B DLS C DALS A DLS A	45.0	3.80 4.80	n.a	1.5 1.8	n.a		(Sinolinks, Linear Staplers)
SINOLINKS	Disposable Linear Stapler	DLS B DLS C DALS A DLS A	60.0	3.80 4.80	n.a	1.5 1.8	n.a		(Sinolinks, Linear Staplers)
SINOLINKS	Disposable Linear Stapler	DLS B DLS C DALS A DLS A	90.0	3.80 4.80	n.a	1.5 1.8	n.a		(Sinolinks, Linear Staplers)
COVIDIEN	DST Series™ TA™ Stapler	TA30V3S TA3035S TA3048S	30.0	2.50 (vascular version) 3.50 4.80	3 2 2	1.0 1.5 2.0	It can be reloaded up to 7 times for a total of 8 firings in a single procedure.	The version with a staple height of 2.5 mm is for vascular applications.	(Medtronic, TA™ Stapler with DST Series™ Technology,) [30]
COVIDIEN	DST Series™ TA™ Stapler	TA4535S TA4548S	45.0	3.50 4.80	2	1.5 2.0	It can be reloaded up to 7 times for a total of 8 firings in a single procedure.	n.a	(Medtronic, TA™ Stapler with DST Series™ Technology) [30]
COVIDIEN	DST Series™ TA™ Stapler	TA6035S TA6048S	60.0	3.50 4.80	2	1.5 2.0	It can be reloaded up to 7 times for a total of 8 firings in a single procedure.	n.a	(Medtronic, TA™ Stapler with DST Series™ Technology) [30]
COVIDIEN	DST Series™ TA™ Stapler	TA9035S TA9048S	90.0	3.50 4.80	2	1.5 2.0	It can be reloaded up to 7 times for a total of 8 firings in a single procedure.	n.a	(Medtronic, TA™ Stapler with DST Series™ Technology) [30]
ETHICON	PROXIMATE® Reloadable Staplers	TX30V TX30B TX30G	30.0	2.50 (vascular version) 3.50 4.80	3 2 2	1.0 1.5 2.0	Max firings: 8	It has applications throughout the alimentary tract and in thoracic surgery. The vascular version has applications in internal tissue, which can easily be compressed to 1mm in thickness and to ligate pulmonary vessels.	(Ethicon, PROXIMATE Reloadable Linear Stapler) [31]
ETHICON	PROXIMATE® Reloadable Staplers	TX60B TX60G	60.0	3.85 4.80	2	1.5 2.0	Max firings: 8	It has applications throughout the alimentary tract and in thoracic surgery.	(Ethicon, PROXIMATE Reloadable Linear Stapler) [31]

Linear Staplers with Bioresorbable Staples

Among all the manufacturers, only Covidien and CooperSurgical (Trumbull, Connecticut, Stati Uniti) include linear staplers mounting absorbable staples. The device of Covidien, TA Premium Polysorb (see Table 3), has the same design and functioning as traditional linear staplers. The available staple length is declared to be 55 mm. Each cartridge is intended to deliver a double staggered row of staples with a closed staple height that can be either 1.5 mm or 5 mm, depending on the staple filament width. The proposed device by CooperSurgical, INSORB®, contains 20 individual, horseshoe-shaped staples with the following features: the staple is nominally 5 mm long × 3.5 mm wide × 0.7 mm thick. The staple is composed of an absorbable copolymer of predominantly polylactide and a lesser component of polyglycolide [32]. A study that compares the use of INSORB® and Subcuticular Skin Closure showed that the two treatment arms did not appear to differ [15,33].

Table 3. Linear staplers with absorbable characteristics.

Brand	Commercial Name	Code	Staple Line Length [mm]	Filament Width [mm]	No. of rows	Closed Staple Height [mm]	Firing Actions	Surgery Application and Relevant Info	Reference
COVIDIEN	TA Premium Polysorb	013501L 013505L	55.0	0.060 0.170	2	1.5 4.3 5.0	n.a	The device uses reabsorbable staples made of Lactomer.	(Covidien Polysorb)

During the search for bioresorbable staplers, a promising device for staple line reinforcement was found. In particular, FOREseal is made of polysaccharidic polyglycuronate biopolymers (highly purified fractions from calcium alginates) originating from seaweeds. It creates an interface between the staples and the organ wall and then forms a hemostatic, healing bioadhesive gel through the release of its calcium ions. The preliminary study on pancreatic parenchyma highlighted that the use of this new device to reinforce the staple line during pancreatectomy was well tolerated [33] and showed promising results for lung staple line reinforcement [34].

In a study conducted in France on six hundred sixty-four patients regarding the dermal suture, it was seen that the suture with semiautomatic stapler is a potential alternative to the one with conventional thread, and regarding patient scar assessment scale score at 3 months or after 18 months, the medium operating time and complication rates were equivalent. Instead, the average suture time and occupational exposure to blood were significantly lower in the stapler [9]. A study conducted on 176 patients confirmed the effectiveness of bioresorbable staples, showing a decreased incidence of composite wound complications with subcuticular staple closure versus traditional staple closure in patients undergoing cesarean section [35].

Another study conducted on 150 patients who underwent distal pancreatectomy affirms that the use of bioresorbable staples, in particular absorbable Lactomer staples for distal pancreatic resection, can be a safe alternative to the standard closure technique [18].

3.2. Linear Cutting Staplers

Linear cutting staplers are employed in open surgery to seal and resect tissues at the same time. The device allocates two or four rows of staggered staples while making a resection in the middle of the rows through the blading element.

Table 4 reports the main characteristics of linear cutting staplers. Panther Medical and Sinolinks offer the same staple line lengths of 60 mm, 80 mm, and 100 mm, while Covidien (part of Medtronic) offers 64 mm, 84 mm, and 104 mm staple line lengths, and Ethicon (part of Johnson & Johnson) proposes different lines of devices varying from the 45 mm staple line length of ENDOPATH® ETS Articulating Linear Cutters to the 102 mm of PROXIMATE® Linear Cutters series. Ethicon ENDOPATH was used in a novel technique that avoids entry to the chest cavity and minimizes the use of electrocautery on the diaphragm [36].

Panther Medical and Sinolinks provide cartridges having the same staple height of 3.8 mm or 4.8 mm, Covidien provides a 2.5 mm option for vascular purposes, while Ethicon offers a variety of sizes covering an interval from 2.6 mm to 4.4 mm depending on the considered series of devices and on the application. As can be seen in Table 3, all producers ensure the stapler reloading a maximum of seven times, resulting in a maximum of eight firings during the same operation. Only the Ethicon brand proposes two devices reaching 12 maximum firings. The Ethicon endo-Surgery Linear Cutter was adopted for gracillis muscle flap splitting in facial reanimation, allowing for a perfect resection of the gracillis muscle [37].

The closed staple height is 1.5 and 2 mm for3.8 and 4.8 mm staple height, respectively, but it reaches a minimum of 1 mm for Ethicon and Covidien cartridges with vascular purposes. Ethicon also offers a 1.8 mm option for medium-thickness tissue application. Covidien linear cutting staplers DST series GIA have been used to prevent blood loss during cesarean delivery [38].

Recently, Meditulip developed a Novel Asymmetrical Linear Stapler (NALS). In comparison with the other liner cutting staplers, on the resected organ side, there is a single row of titanium fasteners located farthest from the endocutter in the stapler device. Thanks to the new design, the stapler can provide tissue for a frozen section biopsy at the true resection margin without the titanium fasteners injuring the tissue [39].

Standard Bariatrics (Cincinnati, OH, USA) has developed a novel linear stapler called Standard Titan to perform a sleeve gastrectomy with a single stapler firing. This device aims to decrease technical and mechanical failure by eliminating junctions along the staple line, decreasing tissue manipulation and operative time while optimizing gastric tissue compression [40].

The performance of the linear cutter stapler (Ethicon Endo-Surgery Linear Cutters) for muscle transferring was tested by Nasir [37]. The results showed a very clear and precise division of the muscle edges; the titanium clips prevent muscle fibril separation, resulting in an easy and safe procedure even after splitting the muscle [37]. The advantages of Ethicon Endo-Surgery Linear Cutters were also reported in a new approach for cutting and closing the pancreas during pancreatojejunostomy with positive results [41] and to perform cervical esophagogastric anastomosis using the triangulating stapling technique [42].

Only one study in the literature reported on the performance of the NALS device developed by Meditulip (Cheongju-si, Chungcheongbuk-do, Republic of Korea). In particular, the NALS stapler was evaluated in the examination of animal lung tissue after the procedure of resection. In comparison with the same typology of staplers, the NALS device showed no squeezing artifact at the resection margin on microscopic examination, and all of the alveolar structures for evaluation were preserved [39]. However, no other information was found on this new linear cutter stapler, and a modification was reported to improve the pathologic evaluation of the true surgical margin after the use of the Ethicon ENDOPATH stapler [43].

Linear cutting staplers are widely used in gastrointestinal surgery. The study carried out by Cheng et al. [44] on 10 patients who underwent laparoscopic total gastrectomy highlighted a safe and effective procedure with the linear cutting stapler [44].

In addition, the efficacy of this type of stapler in a new procedure for anastomosis between the ileum and colon was evaluated and continues to show promising results [39].

Another surgery in which linear cutting staplers are well established consists of laparoscopic sleeve gastrectomy (established standard of care), the most common bariatric procedure. In particular, the stapler is used to resect the lateral portion of the stomach [40]. In comparison with the other two staplers adopted in this surgery, Standard Titan highlighted the mechanical benefits of withstanding higher burst pressure with the notable advantage of single-load functionality. Clearly, the new single staple load eliminates staple line junctions without sacrificing the integrity of staple formation along the staple line [40].

Table 4. Linear cutting stapler characteristics.

Brand	Commercial Name	Code	Staple Line Length [mm]	Staple Height [mm]	No. of Rows	Close Staple Height [mm]	Firing Actions	Surgery Application and Relevant Info	Reference
PANTHER HEALTHCARE	Linear Cutter Stapler	SSAB-60	60.0	3.8 4.8	2	1.5 2.0	It can be reloaded up to 7 times for a total of 8 firings in a single procedure.	n.a	(Panther Healthcare) [45]
PANTHER HEALTHCARE	Linear Cutter Stapler	SSAB-80	80.0	3.8 4.8	2	1.5 2.0			(Panther Healthcare) [45]
PANTHER HEALTHCARE	Linear Cutter Stapler	SSAB-100	100.0	3.8 4.8	2	1.5 2.0			(Panther Healthcare) [45]
SINOLINKS	Disposable Linear Cutter Stapler	DLC B-60	63.0	3.8 4.8	n.a	n.a	n.a	Used for removing and suturing organs in gastrointestinal surgery and general surgery.	(Sinolinks, Disposable Linear Cutter Stapler)
SINOLINKS	Disposable Linear Cutter Stapler	DLC B-80	83.0	3.8 4.8	n.a	n.a			(Sinolinks, Disposable Linear Cutter Stapler)
SINOLINKS	Disposable Linear Cutter Stapler	DLC B-100	103.0	3.8 4.8	n.a	n.a			(Sinolinks, Disposable Linear Cutter Stapler)
COVIDIEN	GIA™ Stapler with DST Series™ Technology	GIA6025S GIA6038S GIA6048S	66.0	2.5 3.8 4.8	4	1.0 1.5 2.0	It can be reloaded for a total of 8 firings.	n.a	(Medtronic, GIA™ Stapler with DST Series™ Technology) [46]
COVIDIEN	GIA™ Stapler with DST Series™ Technology	GIA8038S GIA8048S	86.0	3.8 4.8	4	1.5 2.0	It can be reloaded for a total of 8 firings.	n.a	(Medtronic, GIA™ Stapler with DST Series™ Technology) [46]
COVIDIEN	GIA™ Stapler with DST Series™ Technology	GIA10038S GIA10048S	106.0	3.8 4.8	4	1.5 2.0	It can be reloaded for a total of 8 firings.	n.a	(Medtronic, GIA™ Stapler with DST Series™ Technology) [46]
COVIDIEN	GIA™ Stapler with Tri-Staple™ Technology	(Thin/medium) GIA60TMS GIA60TMC (Medium/thick) GIA60MTS GIA60MTC (Extra thick) GIA60XTS GIA60XTC (Thin/medium) GIA80TMS GIA80TMC (Medium/thick) GIA80MTS GIA80MTC (Extra thick) GIA80XTS GIA80XTC	60 80	2.4 2.7 3.0 3.5 4.0 4.5 5.0	3	n.a	It can be reloaded for a total of 8 firings.	Features: - Stronger staple line leak pressure compared to two-row staplers. - Less stress on tissue during compression and clamping. - A stepped cartridge face and varied-height staples that may provide greater perfusion to the staple line. - Consistent performance over a broad range of tissue thickness.	(Medtronic, GIA™ Stapler with DST Series™ Technology)
ETHICON	Ethicon Surgery Linear Cutters	NTLC55 SR55	55.0	4.4	6	1.5 1.8 2.0	It can be reloaded for a total of 12 firings.	It has applications in gastrointestinal, gynecologic, thoracic, and pediatric surgery for transection, resection, and the creation of anastomoses and can be used with staple line or tissue buttressing materials.	(Ethicon, Linear Cutters) [47]
ETHICON	Ethicon Surgery Linear Cutters	NTLC75 SR75	75.0	4.4	6	1.5 1.8 2.0			(Ethicon, Linear Cutters) [48]

Table 4. *Cont.*

Brand	Commercial Name	Code	Staple Line Length [mm]	Staple Height [mm]	No. of Rows	Close Staple Height [mm]	Firing Actions	Surgery Application and Relevant Info	Reference
ETHICON	PROXIMATE® Linear Cutters	TVC55 TLC55 TCT55	57.0	2.6 3.8 4.5	4	1.0 1.5 2.0	It can be reloaded for a total of 8 firings.	It has applications in gastrointestinal, gynecologic, thoracic, and pediatric surgery for transection, resection, and the creation of anastomoses and can be used with staple line or tissue buttressing materials.	(Ethicon, PROXIMATE® Linear Cutters) [49]
ETHICON	PROXIMATE® Linear Cutters	TLC75 TCD75 TCT75	77.0	3.8 4.2 4.5	4	1.5 1.8 2.0			(Ethicon, PROXIMATE® Linear Cutters) [49]
ETHICON	PROXIMATE® Linear Cutters	TLC10 TCT10	102.0	3.8 4.5	4	1.5 2.0			(Ethicon, PROXIMATE® Linear Cutters) [49]
ETHICON	ENDOPATH® ETS Articulating Linear Cutters (laparoscopic stapler)	ATS45-TR45W ATS45-6R45G ATS45-TR45G	45.0	2.5 3.5 4.1	4 (TR45G) 6 (TR45W or 6R45B)	1.0 1.5 2.0	It can be reloaded for a total of 8 firings.	It has applications in multiple open or minimally invasive general, gynecologic, urologic, thoracic, and pediatric surgical procedures for transection, resection, and/or creation of anastomoses.	(Ethicon, ENDOPATH™ ETS Articulating Linear Cutter) [50]
STANDARD BARIATRICS	Standard Titan	n.a	230	n.a	n.a	1.2 2.2	n.a	The device is designed to help achieve a more consistent and symmetrical performing sleeve gastrectomy. Furthermore, the company may provide a staple line reinforcement with GORE® SEAMGUARD®, a bioabsorbable staple line reinforcement.	(Standard Bariatrics, Standard Titan) [51]

Linear Cutting Staplers with Bioresorbable Staples

Covidien is the only manufacturer to provide a linear cutting stapler having the same design as those described in the previous paragraph but mounting absorbable staples (see Table 5). Poly GIA 75 places two double staggered rows of staples while a knife divides the tissue in between; the staple line length is declared to be 75 mm, and with a staple section width of 0.060 mm, it is possible to reach a closed staple height of 1.5 mm. Covidien also produces Premium Poly CS 57, specifically intended to be used in C-section suturing processes, mounting Polysorb absorbable staples. The design resembles that of linear staplers: the squeezing of a handle activates the firing and cutting mechanism, placing two rows of staples while a knife blade operates a resection between them.

Table 5. Linear cutting staplers with bioresorbable staple characteristics.

Brand	Commercial Name	Code	Staple Line Length [mm]	Filament Width [mm]	No. of Rows	Closed Staple Height [mm]	Firing Actions	Surgery Application and Relevant Info	Reference
COVIDIEN	Poly GIA™ 75	030775	75.0	0.060	2	1.5	n.a	The device uses reabsorbable staples made of Lactomer.	(Medtronic, Poly GIA™ 75)
COVIDIEN	Premium Poly™ CS	05140 05170	57.0	0.140 (thinner tissue) 0.170 (thicker tissue)	2	1.5	n.a	The device uses reabsorbable staples made of Lactomer.	(Medtronic, Premium Poly GIA™ 75)

For over 30 years, the standard of care for hysterotomy in the gravid uterus has been the Premium Poly CS stapler. Premium Poly CS stapler was developed for fetal surgery in the 1980s; the benefits of this stapler included simple use, hemostasis, membrane sealing, and fewer adverse effects upon future reproduction than metallic staplers [52].

Church et al. [53] compared the performance of a bipolar electrocautery device, the LigaSure Impact (Medtronic), and an ultrasonic dissection device, the Harmonic ACE +

7 Shears (Ethicon part of Johnson & Johnson), with the Premium Poly CS Stapler. As a result, the Harmonic ACE could create a hysterotomy that healed comparably to the CS Stapler after repair [53].

3.3. Endo Linear Cutting Staplers

The Endo linear cutting staplers are specialized medical instruments with a crucial role in various surgeries, particularly in gastrointestinal, colorectal, and thoracic procedures. In addition, the use of these devices has been reported for pregnant patients who have undergone surgery, with a low rate of complications reported [54].

The device features an adjustable articulating element controlled by a lever located on the upper part of the shaft. Most manufacturers allow for a 22° range of movement in one direction and 23° in the other, providing a total articulation movement of 45°.

In this work, Table 6 outlines the main characteristics of the Endo linear staplers. In detail, Panther Medical, Sinolinks, and Covidien offer a consistent triplet of shaft length options, ranging from 6 cm to 26 cm. However, Ethicon stands out with shaft length options of 34 cm and 44 cm and an additional 32 cm alternative specifically designed for vascular applications. On the other hand, the staple line length varies depending on the cartridge in use, with the standard options being 30, 45, and 60 mm, except for Sinolinks, which ensures staple line lengths of 31, 46, and 61 mm. Ethicon's PVE35A model offers a specific staple line length of 35 mm, tailored for vascular applications, as evidenced above, which explains the difference in comparison to other models. Except for Ethicon, which presents a range of values from 2.6 mm for the vascular cartridge to 4.2 mm for the thick tissue cartridge, the open staple height is relatively standardized among manufacturers. The Endo GIA linear cutting stapler is reported to be used in vascular applications [55].

Each manufacturer employs a varying number of staple rows for their devices. Sinolinks and Ethicon use three rows on each side of the cutting line, whereas Panther Medical and Covidien opt for two rows on each side of the blade. This results in a total of four lines. To prioritize safety and prevent blood leakage while ensuring successful anastomosis, the staple line length exceeds the cutting line length by at least 5 mm. All the devices under consideration guarantee a 45° degree of reached articulation.

Endoscopic linear cutting staplers have garnered considerable attention and discussion within the scientific community due to their remarkable applications across various types of endoscopic surgeries. In particular, The Endo GIA™ linear stapler has proven to be highly effective and superior to its competitors, specifically in gastric laparoscopy surgery, surpassing products like ETS flex™. This superiority is attributed to its outstanding materials, durability, and reusability after multiple surgeries, resulting in improved clinical outcomes. Notably, it exhibits exceptional efficacy in reducing excessive blood loss and minimizing human errors during procedures [4].

Endoscopic staplers are adopted for the treatment of patients with advanced lung emphysema. In particular, Akil et al. [56] evaluated the surgical performance of a traditional Endo linear stapler compared to a powered stapling system, reporting no difference in the amount of postoperative air leakage, the consecutive length of chest tube drainage, and the time of surgery [56].

In addition to this, some authors have further highlighted the innovative applications of Endo GIA™ linear staplers in pancreaticoduodenectomy and total pancreatectomy, addressing the challenges faced by many surgeons in these complex surgeries. Its use in these procedures offers significant benefits, including reduced bleeding, enhanced surgical precision, shorter operative times, and ultimately, decreased postoperative morbidity and mortality rates [57].

With regard to other manufacturers, Ethicon offers a range of device instruments specifically designed for minimally invasive surgical procedures, including gastrointestinal, vascular, and thoracic procedures, in comparison to Covidien products. In particular, recent studies have emphasized the remarkable efficacy of powered vascular staplers (PVS) in intricate surgical procedures, such as thoracoscopic anatomical surgery, where endoscopic staplers play a critical role in transecting vessels, bronchi, and lung tissue. The ECHELON FLEX™ (Ethicon PSE45A) powered articulating linear instrument stands out as the preferred choice for closing targeted bronchi, consistently leading to highly satisfactory clinical outcomes. Its advantage becomes evident in the context of limited surgical space, making it significantly advantageous over other commercially available powered vascular staplers. The ECHELON FLEX™ has proven its superiority and ease of use, offering equivalent levels of effectiveness and security. By integrating the ECHELON FLEX™ into surgical procedures, the overall complexity of the operation is reduced, and the potential for intraoperative complications is minimized [58,59].

Notably, the use of Ethicon powered staplers has been associated with lower rates of transfusion, as evident in comparison with Medtronic manual staplers. Moreover, recent studies have compellingly supported the use of the ECHELON FLEX™ for pulmonary artery stapling, showcasing its promise in preventing tissue damage and intraoperative bleeding during pulmonary artery transection procedures [60].

On the other hand, some authors' reports have compellingly showcased the efficacy of the Panther and Sinolink families of linear staplers in laparoscopic total gastrectomy for cancer treatments. Among these breakthroughs, the intracorporeal construction of the anastomosis stands out, where endoscopic linear staplers have emerged as the preferred reconstruction method among esteemed surgeons worldwide, surpassing circular staplers in popularity and reliability. The overwhelming preference for linear stapled anastomosis over circular stapled anastomosis stems from its numerous advantages and simplified execution. Notably, the linear approach is widely regarded as being easier to perform, making it an attractive choice for surgeons seeking optimal outcomes and reduced surgical complexity [61,62].

Table 6. Endo linear cutting stapler characteristics.

Brand	Commercial Name	Code	Shaft Length [cm]	Staple Line Length [mm]	Staple Height [mm]	No. of Rows	Close Staple Height [mm]	Firing Actions	Surgery Application and Relevant Info	Reference
PANTHER HEALTHCARE	Universal Endo Linear Cutter	CEAC 30 45 60	6 16 26	30 45 60	2.5 3.5 4.2 4.8 5.0	4	n.a	It can be reloaded for a total of 25 firings.		(Panther Healthcare) [63]
PANTHER HEALTHCARE	Endo Linear Stapler and P2G Cartridges	CADF-30T CADF-30N CADF-45T CADF-45N CADF-45S CADF-45D CADF-45R CADF-60T CADF-60N CADF-60N CADF-60S CADF-60D CADF-60R	6 16 26	30 45 60	2.5 3.5 4.2 4.8 5.0	n.a	n.a	n.a	Features: Dog bone-shaped staple buckets enhance tissue stability and provide improved staple formation in thick tissue.	(Panther Healthcare) [63]
SINOLINKS	Disposable Endoscopic Linear Cutter, GEN I and GEN II	DEC A/B 60 160 260	26	31 46 61	2 2.5 3.5 4.8	6	1.5 1.9	It can be reloaded for a total of 25 firings.	It has applications in thoracic surgery, abdominal surgery, pediatric surgery, and gynecologic endoscopy procedures for transection, resection, and/or creation of anastomoses.	(Sinolinks, Endoscopic linear stapler)

Table 6. *Cont.*

Brand	Commercial Name	Code	Shaft Length [cm]	Staple Line Length [mm]	Staple Height [mm]	No. of Rows	Close Staple Height [mm]	Firing Actions	Surgery Application and Relevant Info	Reference
COVIDIEN	Multifire Endo TA™ 30 Staplers and Reloads	n.a	n.a	30	2.5 (vascular) 3.5	3	n.a	It can be reloaded 7 times for a total of 8 firings.	n.a	(Medtronic, Multifire Endo TA™) [64]
COVIDIEN	Endo GIA™ Universal Staplers with Tri-Staple	SIG30AVM SIG30AVT SIG45AXT SIG60AXT	n.a	30 45 60	2.0 2.5 3.0 3.5 4.0	3	n.a	n.a	n.a	(Medtronic, Endo GIA™ Ultra Universal Staplers) [64]
COVIDIEN	Endo Gia™ Ultra Universal Stapler	EGIAUSHT EGIAUSTD EGIAUXL	6 16 26	30 45 60	2.0 2.5 3.5 4.8	6	0.75 0.88 1.5 2.25 3.0	n.a	n.a	(Medtronic, Endo GIA™ Ultra Universal Staplers) [65]
ETHICON	ECHELON FLEX™ ENDOPATH™ Staplers	EC45A EC45AL EC60A LONG60A	34 44	45 60	2.6 3.6 3.8 4.1 4.2	6	1 1.5 1.8 2 2.3	n.a	It has applications in multiple open or minimally invasive general, gynecologic, urologic, thoracic, and pediatric surgical procedures. It can be used with tissue buttressing materials. The instrument may also be used for transection and resection of liver parenchyma (hepatic vasculature and biliary structures), pancreas, kidney, and spleen.	(Ethicon, Echelon Flex Endopath) [66]

3.4. Circular Staplers

Specialists use them for end-to-side, end-to-end, and side-to-side anastomoses. The applications of circular staplers involve the anastomosis of the proximal colon with the distal colon during colostomy, laparoscopic distal gastrectomy, and gastrojejunal anastomosis in Roux-en-Y gastric bypass [67–73]. Circular staplers also play an increasingly important role in vascular anastomoses [74]. The instrument is composed of a detachable anvil and a handling part, the last comprising a shaft that allocates the stapling cartridge and cutting element, a wing nut for opening and closing, and a handle to perform the firing and cutting operation [75]. This class of devices' cartridges contain a circular knife and a crown of staples of different heights that are available in a range of diameters to adapt to the dimension of the lumen to suture.

Table 7 shows the principal characteristics of circular staplers, sorting devices according to their producers and identifying them both through their commercial name and through their univocal product code. Panther Medical offers a device with a 26 cm shaft and cartridges with 21 mm, 24 mm, 26 mm, 29 mm, 32 mm, and 34 mm diameters. No information on the shaft length has been found for Sinolinks. Still, it is known that their staplers are available in 17 mm, 19 mm, 21 mm, 23 mm, 25 mm, and 27 mm diameters. Covidien offers 22 cm and 35 cm shafts with diameters of 21 mm, 25 mm, 28 mm, 31 mm, and 33 mm. Finally, Ethicon circular staplers have 26 and 37 cm shafts with 21 mm, 25 mm, 29 mm, and 33 mm diameters. As can be seen in Table 7, all manufacturers, except for Sinolinks, for which data were not available, provide cartridges allocating two rows of staples whose height depends on the chosen diameter; only Covidien uses its technology

Tri-Staple, with three rows of different height staples. Circular staplers' cartridges contain a circular knife and a crown of staples of different heights, and they are available in a range of diameters to adapt to the dimension of the lumen. Staple height varies in a wide range depending on the considered manufacturer, going from 3 mm of EEA™ circular stapler with Tri-Staple™ technology to 5.5 mm of Ethicon circular staplers. Close staple heights also vary in the data, including a range that goes from 1.4 mm for Panther Medical to 3mm for Ethicon circular staplers. Ethicon circular staplers with product codes beginning with CDHA and ECS-A provide the maximum closed staple height, indicating they are specifically intended for use in deeper, thicker bowels.

Despite the widespread use of staplers, surgeons are still in search of the most suitable technique to perform surgery. Circular staplers are typically used in colorectal anastomosis, gastrojejunal anastomosis, and other procedures. In the meta-analysis of Milone et al. [73], the literature regarding the use of circular and linear stapler techniques for gastrojejunostomy was analyzed. The results showed that by analyzing the overall anastomotic complications, there was a significantly lower rate of complications in the linear stapler technique compared to the circular stapler one. Other studies in the literature showed similar results. Edholm et al. [76] found that linear staplers, unlike circular staplers, are associated with short operative times, reduced wound infection risks, and shorter lengths of stay.

Mazaki et al. [77] compared the pressure resistance of triple-row circular stapler (EEA™ circular stapler with Tri-Staple™ technology, 28 mm Medium/Thick, Covidien) and double-row circular stapler (EEA™ circular stapler with DST series™ technology, 28 mm, 4.8 mm, Covidien) in a porcine model, reporting that the pressure resistance of the triple-row circular stapler is higher compared to DCS, suggesting that the triple-row circular stapler may reduce the rate of anastomotic leakage [77].

Strassner et al. (2023) [78] evaluated the performance of the Tri-Staple technology (Medtronic) with two-row staplers (Ethicon), comparing them in terms of perfusion, withstanding tension and intraluminal pressure, and reduced damage to the staple line during removal. The results of the tests show that Tri-Staple technology shows a median leak pressure that was 73% higher, better perfusion preservation, a 20% higher failure force, and a median removal force 78% lower than the double-row stapler [78].

Table 7. Circular stapler characteristics.

Brand	Commercial Name	Code	Shaft Length [cm]	Diameter [mm]	Staple Height [mm]	No. of Rows	Close Staple Height [mm]	Staple Quantity	Surgery Application and Relevant Info	Reference
PANTHER HEALTHCARE	Disposable Circular Stapler	FCSME21 FCSME24 FCSME26 FCSME29 FCSME32 FCSLWAF21 FCSLWAF25 FCSLWAF29 FCSLWAF32 FCSLWAF33 (Extra Long)	26.0	21.0 24.0 26.0 29.0 32.0 21.0 25.0 29.0 32.0 33.0	4.5 4.8 5.0	2	1.4-2.2 (range)	n.a	n.a	(Panther Healthcare, Circular Stapler) [79]
SINOLINKS	Disposable Circular Stapler	DCS H 17 DCS H 19	n.a	17.0 19.0	4.5	n.a	2.0 2.2	n.a	It has applications in gastrointestinal surgery. Used for the whole digestive tract for end-to-end, end-to-side, and side-to-side anastomoses.	(Sinolinks, Disposable Circular Stapler)
SINOLINKS	Disposable Circular Stapler	DCS A/E/F21 DCS A/E/F23 DCS A/E/F25 DCS A/E/F27	n.a	21.0 23.0 25.0 27.0	4.8 5.2	n.a	2.0 2.2	18 18 20 22		(Sinolinks, Disposable Circular Stapler)
SINOLINKS	Disposable Circular Stapler	DCS A/E/F29 DCS A/E/F31 DCS A/E/F33	n.a	29.0 31.0 33.0	4.8 5.2	n.a	2.0 2.2	24 28 30		(Sinolinks, Disposable Circular Stapler)

Table 7. *Cont.*

Brand	Commercial Name	Code	Shaft Length [cm]	Diameter [mm]	Staple Height [mm]	No. of Rows	Close Staple Height [mm]	Staple Quantity	Surgery Application and Relevant Info	Reference
COVIDIEN	EEA™ Stapler with DST Series™ Technology	EEA21 EEA25 EEA28	22.0 35.0	21.0 25.0 28.0	3.5 4.8	2	n.a	18 22 26 30 32	It has applications in bariatric, colorectal, and general surgery.	(Medtronic, EEA™ Circular Stapler with DST™ Series Technology) [80]
COVIDIEN	EEA™ Stapler with DST Series™ Technology	EEA31 EEA33	22.0 35.0	31.0 33.0	4.8	2	n.a			(Medtronic, EEA™ Circular Stapler with DST™ Series Technology) [80]
COVIDIEN	EEA™ Stapler with Tri-Staple Technology	TRIEEA28MT TRIEEA28XT TRIEEAXL28MT TRIEEAXL28XT TRIEEA31MT TRIEEA31XT TRIEEAXL31MT TRIEEAXL31XT	n.a	28.0 31.0	MT 3.0 3.5 4.0 XT 4.0 4.5 5.0 XL 3.0 3.5 4.0	3	n.a	n.a	n.a	(Medtronic, Suturatrice Circolare EEA™ con Tri-Staple™ Technology) [81]
ETHICON	Ethicon Circular Stapler	CDH21A CDH25A CDH29A CDH33A	26.0	21.0 25.0 29.0 33.0	5.5	2	1.5 3.0	16	n.a	(Ethicon, Circular Stapler) [82]
ETHICON	Ethicon Circular Stapler	ECS21A ECS25A ECS29A ECS33A	37.0	21.0 25.0 29.0 33.0	5.5	2	1.5 3.0	16 20 24 28	n.a	(Ethicon, Circular Stapler) [83]
ETHICON	ETHICON™ Circular Stapler	ECS21B ECS25B ECS29B ECS33B	37.0	21.0 25.0 29.0 33.0	5.2	2	1.5 2.2	16 20 24 28	It has applications in surgery fields such as bariatric, colorectal, and thoracic surgeries to perform esophagectomy, gastric bypass, left colectomy, lower anterior resection, and sigmoidectomy procedures.	(Ethicon, Circular Stapler) [84]

Over the years, researchers and surgeon specialists performed different studies to assess which type of stapler was the most suitable for different surgery procedures. Gastrojejunostomy is one of the most analyzed techniques for anastomosis. Typically, two different types of staplers' outcomes are compared, and the authors consider linear and circular staplers' techniques for the outcome comparison [3,68–73,76].

Other studies analyze the safety and efficacy of laparoscopy gastrectomy in patients with gastric cancer using linear stapler and circular stapler techniques [85]. Circular staplers were also used for restorative surgery for rectal cancer [86].

A variant of this class of device is hemorrhoidal staplers, which work exactly as circular staplers but have different diameters and employ different cartridges, ensuring a lower closed staple height. Table 8 below displays the different manufacturers that offer staplers for this specific purpose. The disposable hemorrhoidal stapler by Sinolinks has diameters of 32 mm and 34 mm; Covidien has the EEA™ hemorrhoid and prolapse stapler with DST Series™ technology that has a diameter of 33 mm, the same as Ethicon PROXIMATE® PPH3 Hemorrhoidal Circular Stapler Set; and Golden Stapler has the PPHD Single-Use Hemorrhoidal Circular Stapler line, with diameters of 32 mm, 34 mm, and 36 mm. The staple height is different in each of the four staplers: 3.8 mm for Sinolinks staples, 3.5 mm and 4.8 mm for Covidien staples, 4 mm for Ethicon, and finally, 4.3 and 4.5 for Golden Stapler. The number of rows of staggered staples, as the closed staple height, is also listed in Table 8.

Table 8. Hemorrhoidal stapler characteristics.

Brand	Commercial Name	Code	Diameter [mm]	Staple Height [mm]	No. of Rows	Closed Staple Height [mm]	Staple Quantity	Surgery Application and Relevant Info	Reference
SINOLINKS	Disposable Hemorrhoidal Stapler	DHS B-32 DHS B-34	32.0 34.0	4	2	1	32 34	It has applications in hemorrhoid surgery for the selective resection of the mucosa along the dentate line.	(Sinolinks, Disposable Circular Stapler) [87]
SINOLINKS	Disposable Hemorrhoidal Stapler	DHS C-32 DHS C-34	32.0 34.0	4	2	1	32 34		
COVIDIEN	EEA™ Hemorrhoid and Prolapse Stapler with DST Series™ Technology	HEM3335	33.0	3.5 4.8	4	n.a	n.a	It has applications in hemorrhoid surgery.	(Medtronic, EEA™ Hemorrhoid and Prolapse Stapler Set) [88]
ETHICON	PROXIMATE® PPH Hemorrhoidal Circular Stapler Set	PPH03	33.0	4.0	2	0.75 1.50	28	It has applications in hemorrhoid surgery.	(Ethicon, PROXIMATE™ PPH Hemorrhoidal Circular Stapler Set) [89]
GOLDEN STAPLER	Single-Use Hemorrhoidal Circular Stapler	PPHD32 PPHD34 PPHD36	32.5 34.5 36.5	4.3 4.3 4.5	2	0.8 2.0	32 34 36	It has applications in hemorrhoid surgery.	(Golden Stapler, Single Use Hemorrhoidal Circular Stapler) [79]

3.5. *Powered Stapling System*

Both linear and circular powered stapling systems are commercially available (see Table 9). For linear systems, today, Ethicon and Covidien are the only manufacturers on the market offering powered surgical staplers. Ethicon has two powered stapling systems, one being the ECHELON CIRCULAR™ Powered Stapler (ECPS) already mentioned in the previous section. The other one is the ECHELON FLEX™ Powered Vascular Stapler (PVS) (Ethicon, Echelon Flex Powered Vascular Stapler) [90].

The PVS has a powered actuation, meaning that the firing mechanism is no more controlled by the swing of a knob or the compression of a spring but through a battery providing power to an electro-mechanical system maneuvering the cutting element if present and the cartridge. The stapler has a curved anvil tip, an introducer-like cartridge body tip, and active articulation with high articulation angles to facilitate access in surgical procedures with small surgical spaces or difficult-to-reach vessels.

Medtronic Covidien Signia™ Stapling System (SIG) is relatively more complex. Not only does it have powered actuation, but the articulation, opening, and closing of the end effector are also powered through electrical motors. The compression of the tissue is sensed thanks to a strain-gage transducer located on the shaft, providing indications to the surgeon through a display.

Regarding circular powered stapling solutions, Ethicon produces the ECHELON CIRCULAR™ Powered Stapler (ECPS) (Ethicon, Echelon Circular Powered Vascular Stapler) [91].

This represents the powered variant of the already-discussed ETHICON™ Circular Stapler product and shares most of its technical specifications, such as lumen diameter (available in four sizes), staple rows, open staple leg length, and closed staple height. The iDrive™ staplers provided by Covidien showed superior performance in terms of access, visibility, and ease of placement in the deep pelvis [92].

Both the ECHELON FLEX™ Powered Vascular Stapler (PVS) and the Covidien Signia™ Stapling System (SIG) have been extensively validated by the scientific community. In particular, SIG was tested for video-assisted thoracic surgery by Shimada et al. [93], which also introduced a new procedure, named "sliding technique", to optimize safe surgical margins with minimal stapler movement.

The advantages of PVS in comparison to manual endoscopy are highlighted in Roy et al. [94] for laparoscopic bariatric surgery, in Park et al. [95] for thoracoscopic lobectomy lung cancer, and video-assisted thoracic surgery lobectomy in Miller et al. [59].

In comparison to the non-powered alternatives, these studies showed, respectively, a decrease of 47% (3.05% to 1.61%) in terms of hemostasis-related complications, a decrease of 56% in terms of intraoperative blood loss, and a decrease of 47% in terms of bleeding complications. The PVS device has also been applied for scientific investigations on a novel physiologic lung model by Eckert et al. [96].

Some direct comparisons of PVS and SIG have been performed by Rawlins et al. [97] in laparoscopic sleeve gastrectomy. The authors report a lower observed incidence of hemostasis-related complications during the surgical admission in the ECHELON FLEX™ group in comparison with the SIG group (3 events vs. 11 events).

Regarding the ECHELON CIRCULAR™ Powered Stapler (ECPS), this has been proven to reduce the risk of anastomotic leakage rates in left-sided colorectal anastomosis from 11.8% to 1.7% in a single institution clinical study [98]. Postoperative anastomotic leaks were experienced in 2.4% of subjects in a cohort study of left-sided colorectal surgery performed by Herzig et al. [99]. Similarly, it has been successfully benchmarked against manual operation for left-sided colorectal resections by Pollack et al. [100] and Sylla et al. [101].

Table 9. Powered stapler characteristics.

Brand	Commercial Name	Code	Dimension [mm]	Staple Height [mm]	No. of Rows	Closed Staple Height [mm]	Firing Actions	Surgery Application and Relevant Info	Reference
COVIDIEN	Signia™ Stapling Platform	n.a	30 30 45	4.5	n.a.	n.a.	n.a	It adapts to different surgical conditions, properly adjusting the drive speed according to force measurements.	(Medtronic, Signia™ Stapling Platform) [102]
COVIDIEN	Signia™ Stapling Platform Small Diameter (8 mm shaft)	SIGSDS30CTV SIGSDS30CTVT SIGSDS45CTVT	30 30 45	n.a	n.a	n.a	n.a	Compatibility with all Medtronic handles.	(Medtronic, Signia™ Stapling Platform) [102]
ETHICON	FLEX™ Powered Vascular Stapler	PVE35A VASECR35 (Only for vascular/ thin tissue)	32	5.2	4	1.5 mm–2.2 mm	It can be reloaded for a total of 12 firings.	It has applications in surgery fields such as general, gynecologic, pediatric, thoracic, and urologic to perform lung resection, nephrectomy, appendectomy, splenectomy, and oophorectomy procedures, as well as for vessel transection and stapling.	(Ethicon, Echelon Flex Powered Vascular Stapler) [90]
ETHICON	CIRCULAR™ Powered Vascular Stapler	CDH23P, CDH25P, CDH29P, CDH31P	23 mm, 25 mm, 29 mm, 31 mm	5.2	2	1.5 mm–2.2 mm	n.a	It has applications in surgery fields such as colorectal, gastric, and thoracic to perform colorectal resections, gastrectomies, gastric bypass, and esophagectomies procedures.	(Ethicon, Echelon Circular Powered Vascular Stapler) [91]
COVIDIEN	iDrive™	n.a	n.a	n.a	n.a	n.a	n.a	The iDrive Ultra powered stapling device, with Endo GIA stapling reloads, provides precision in surgical stapling, enabling surgeons to position and keep the stapler exactly where they intend.	(Covidien, iDrive™)

4. Conclusions

Stapling devices have gained widespread adoption in various surgical fields, demonstrating excellent results in terms of surgical outcomes and reduced operation time. Manufacturers and the existing literature provide valuable information on the specific applications of each stapling device, with a range of staple cartridges tailored to different surgical targets. The variety of staple sizes, designed to accommodate diverse tissue thicknesses and lumen diameters, surgical spaces, ensures effective hemostasis while minimizing ischemia and tissue damage. However, it is important to acknowledge that stapling can disrupt the structural continuity of tissues, introducing stress and strain that may not be entirely physiological but still lead to favorable outcomes. Despite this, only a limited number of articles have explored the strength of anastomotic techniques involving stapling devices. It is important to highlight that staple configurations are under constant improvement, and the experimental tests to evaluate their performance are performed by different research groups, as reported in [78].

Therefore, further studies should focus on understanding the impact of adopting specific stapling devices, also in comparison with manual or mixed operations as documented in [2] regarding suture resistances, for different tissues and explore the potential influence of staple height and the number of staple rows on the overall surgical outcome. By undertaking such investigations, we can better optimize the use of stapling devices, tailor their applications to specific tissues, and potentially enhance the overall success and safety of surgical procedures. This research will contribute to the advancement of surgical techniques and the improvement of patient outcomes in the future.

This review aims to gather information from the literature published in previous years regarding surgical stapling devices. Furthermore, specific sections are dedicated to the use of bioresorbable staples to highlight where bioresorbable materials are used in this specific field of surgery. The use of bioresorbable devices in medicine has been increasing in the last few years, both in clinical and research settings. However, few devices adopt the usage of bioresorbable staples and only for specific fields of surgery. The collected information indicates a growing interest in new stapling devices using bioresorbable staples.

While this suggests the potential future of bioresorbable staples, their applicability across all surgical fields might be limited only to skin closure. Indeed, non-bioresorbable staples will remain the gold standard for applications that require permanent closure of internal organs, such as gastrointestinal procedures, abdominal surgery, gynecology, and vascular surgery.

Author Contributions: V.B.: Investigation, Writing—Original Draft Preparation, Writing—Review and Editing; J.B.: Investigation, Writing—Original Draft Preparation; M.R.R.: Investigation, Writing—Review and Editing, Writing—Original Draft Preparation, Formal analysis; M.C.: Investigation, Writing—Original Draft Preparation, Writing—Review and Editing, Data Curation, Formal analysis; O.G.R.: Investigation, Writing—Original Draft Preparation, Writing—Review and Editing, Formal analysis; C.S.: Conceptualization, Writing—Review and Editing, Methodology, Supervision, Funding acquisition, Resources; N.M.P.: Conceptualization, Methodology, Writing—Review and Editing, Supervision. All authors have read and agreed to the published version of the manuscript.

Funding: This research received no external funding.

Acknowledgments: The authors would like to thank Giorgio Tropiano for their precious help in the first draft of the paper and Jacobacci & Partners (Turin) for the help in the patent landscape.

Conflicts of Interest: The authors declare no conflicts of interest.

References

1. Yang, J.; Soltz, M.; Russell, H.; Beres, J.; Zhao, J.; Liao, D.; Gregersen, H. Surface Deformation Analysis of End-to-End Stapled Intestinal Anastomosis. *Surg. Innov.* **2012**, *19*, 281–287. [CrossRef]
2. Lauro, E.; Corridori, I.; Luciani, L.; Di Leo, A.; Sartori, A.; Andreuccetti, J.; Trojan, D.; Scudo, G.; Motta, A.; Pugno, N.M. Stapled Fascial Suture: Ex Vivo Modeling and Clinical Implications. *Surg. Endosc.* **2022**, *36*, 8797–8806. [CrossRef] [PubMed]

3. Penna, M.; Markar, S.R.; Venkat-Raman, V.; Karthikesalingam, A.; Hashemi, M. Linear-Stapled Versus Circular-Stapled Laparoscopic Gastrojejunal Anastomosis in Morbid Obesity: Meta-Analysis. *Surg. Laparosc. Endosc. Percutaneous Tech.* **2012**, *22*, 95–101. [CrossRef] [PubMed]
4. Champion, J.K.; Williams, M.D. Prospective Randomized Comparison of Linear Staplers during Laparoscopic Roux-En-Y Gastric Bypass. *Obes. Surg.* **2003**, *13*, 855–859. [CrossRef] [PubMed]
5. Nakanishi, R.; Fujimoto, Y.; Sugiyama, M.; Hisamatsu, Y.; Nakanoko, T.; Ando, K.; Ota, M.; Kimura, Y.; Oki, E.; Yoshizumi, T. Clinical Impact of the Triple-Layered Circular Stapler for Reducing the Anastomotic Leakage in Rectal Cancer Surgery: Porcine Model and Multicenter Retrospective Cohort Analysis. *Ann. Gastroenterol. Surg.* **2022**, *6*, 256–264. [CrossRef] [PubMed]
6. Lee, S.; Lee, H.; Song, J.H.; Choi, S.; Cho, M.; Son, T.; Kim, H.-I.; Hyung, W.J. Intracorporeal Esophagojejunostomy Using a Linear Stapler in Laparoscopic Total Gastrectomy: Comparison with Circular Stapling Technique. *BMC Surg.* **2020**, *20*, 100. [CrossRef]
7. Lim, J.J.B.; Erdman, A.G. A Review of Mechanism Used in Laparoscopic Surgical Instruments. *Mech. Mach. Theory* **2003**, *38*, 1133–1147. [CrossRef]
8. Von Maszewski, M.; Sucher, J.F.; MacFadyen, B.V. Laparoscopic Instrumentation: Linear Cutters, Clip Appliers, and Staplers. *Semin. Laparosc. Surg.* **2001**, *8*, 69–76. [CrossRef]
9. Giaccaglia, V.; Antonelli, M.S.; Addario Chieco, P.; Cocorullo, G.; Cavallini, M.; Gulotta, G. Technical Characteristics Can Make the Difference in a Surgical Linear Stapler. or Not? *J. Surg. Res.* **2015**, *197*, 101–106. [CrossRef]
10. Yao, D.B.; Wu, S.D. Application of Stapling Devices in Liver Surgery: Current Status and Future Prospects. *World J. Gastroenterol.* **2016**, *22*, 7091–7098. [CrossRef]
11. Tajirian, A.L.; Goldberg, D.J. A Review of Sutures and Other Skin Closure Materials. *J. Cosmet. Laser Ther.* **2010**, *12*, 296–302. [CrossRef]
12. Duteille, F.; Rouif, M.; Alfandari, B.; Andreoletti, J.B.; Sinna, R.; Laurent, B.; Perrot, P. Reduction of Skin Closure Time without Loss of Healing Quality: A Multicenter Prospective Study in 100 Patients Comparing the Use of Insorb Absorbable Staples with Absorbable Thread for Dermal Suture. *Surg. Innov.* **2013**, *20*, 70–73. [CrossRef]
13. Partecke, L.I.; Kessler, W.; Von Bernstorff, W.; Diedrich, S.; Heidecke, C.D.; Patrzyk, M. Laparoscopic Appendectomy Using a Single Polymeric Clip to Close the Appendicular Stump. *Langenbecks Arch. Surg.* **2010**, *395*, 1077–1082. [CrossRef] [PubMed]
14. Kuppersmith, R.B.; Atkins, J.H.; Tami, T.A. The Use of Bioresorbable Staples for Mucoperichondrial Flap Coaptation in Septoplasty. *Otolaryngol. Head. Neck Surg.* **2009**, *140*, 599–600. [CrossRef] [PubMed]
15. Malard, O.; Duteille, F.; Darnis, E.; Espitalier, F.; Perrot, P.; Ferron, C.; Planche, L.; Hardouin, J.B.; Tessier, P.; Bellanger, M.; et al. A Novel Absorbable Stapler Provides Patient-Reported Outcomes and Cost-Effectiveness Noninferior to Subcuticular Skin Closure: A Prospective, Single-Blind, Randomized Clinical Trial. *Plast. Reconstr. Surg.* **2020**, 777E–789E. [CrossRef] [PubMed]
16. Farkas, G.; Leindler, L.; Farkas, G. Safe Closure Technique for Distal Pancreatic Resection. *Langenbecks Arch. Surg.* **2005**, *390*, 29–31. [CrossRef] [PubMed]
17. Hirashima, T.; Eto, T.; DenBesten, L. Lactomer Copolymer Absorbable Staples in Gastrointestinal Surgery. *Am. J. Surg.* **1985**, *150*, 381–385. [CrossRef] [PubMed]
18. Farkas, G. PolysorbR (an Absorbable Lactomer) Staples, a Safe Closure Technique for Distal Pancreatic Resection. *World J. Gastroenterol.* **2014**, *20*, 17185. [CrossRef] [PubMed]
19. Amano, H.; Miyake, K.; Hinoki, A.; Yokota, K.; Kinoshita, F.; Nakazawa, A.; Tanaka, Y.; Seto, Y.; Uchida, H. Novel Zinc Alloys for Biodegradable Surgical Staples. *World J. Clin. Cases* **2020**, *8*, 504–516. [CrossRef]
20. Amano, H.; Hanada, K.; Hinoki, A.; Tainaka, T.; Shirota, C.; Sumida, W.; Yokota, K.; Murase, N.; Oshima, K.; Chiba, K.; et al. Biodegradable Surgical Staple Composed of Magnesium Alloy. *Sci. Rep.* **2019**, *9*, 14671. [CrossRef]
21. Cole, B.J.; Sayegh, E.T.; Yanke, A.B.; Chalmers, P.N.; Frank, R.M. Fixation of Soft Tissue to Bone. *J. Am. Acad. Orthop. Surg.* **2016**, *24*, 83–95. [CrossRef] [PubMed]
22. Barca, F.; Busa, R. Resorbable Poly-l-Lactic Acid Mini-Staples for the Fixation of Akin Osteotomies. *J. Foot Ankle Surg.* **1997**, *36*, 106–111. [CrossRef] [PubMed]
23. Rethnam, U.; Kuiper, J.; Makwana, N. Mechanical Characteristics of Three Staples Commonly Used in Foot Surgery. *J. Foot Ankle Res.* **2009**, *2*, 5. [CrossRef] [PubMed]
24. Oberlander, M.A.; Chisar, M.A. Meniscal Repair Using the Polysorb Meniscal Stapler XLS. *Arthrosc. J. Arthrosc. Relat. Surg.* **2005**, *21*, 1148.e1–1148.e5. [CrossRef] [PubMed]
25. McMillan, S.; Ford, E. Management of Rotator Cuff Defects After Calcific Tendinopathy Debridement Using a Bioinductive Collagen Implant. *Arthrosc. Tech.* **2016**, *5*, e1373–e1379. [CrossRef]
26. Gong, J.; Guo, Z.; Li, Y.; Gu, L.; Zhu, W.; Li, J.; Li, N. Stapled vs Hand Suture Closure of Loop Ileostomy: A Meta-Analysis. *Color. Dis.* **2013**, *15*, e561–e568. [CrossRef] [PubMed]
27. Choy, P.Y.G.; Bissett, I.P.; Docherty, J.G.; Parry, B.R.; Merrie, A. Stapled versus Handsewn Methods for Ileocolic Anastomoses. In *Cochrane Database of Systematic Reviews*; Bissett, I.P., Ed.; John Wiley & Sons, Ltd.: Chichester, UK, 2007.
28. Kano, M.; Hanari, N.; Gunji, H.; Hayano, K.; Hayashi, H.; Matsubara, H. Is "Functional End-to-End Anastomosis" Really Functional? A Review of the Literature on Stapled Anastomosis Using Linear Staplers. *Surg. Today* **2017**, *47*, 1–7. [CrossRef]
29. Panther Healthcare Canada, Linear Stapler and Cartridges. Available online: https://pantherhealthcarecanada.com/products/open-linear-stapler/ (accessed on 19 July 2023).

30. Medtronic, TA^{TM} Sta-Pler with DST Se-Ries^{TM} Technology. Available online: https://www.medtronic.com/covidien/en-us/support/products/surgical-stapling/dst-series-ta-single-use-staplers-and-reloads.html (accessed on 19 July 2023).
31. Ethicon, PROXIMATE Reloadable Linear Stapler. Available online: https://www.jnjmedtech.com/en-us/product/proximate-tx-reloadable-linear-stapler (accessed on 19 July 2023).
32. Pineros-Fernandez, A.; Salopek, L.S.; Rodeheaver, P.F.; Drake, D.B.; Edlich, R.; Rodeheaver, G.T. A Revolutionary Advance in Skin Closure Compared to Current Methods. *J. Long. Term. Eff. Med. Implant.* **2006**, *16*, 19–27. [CrossRef]
33. Hornez, E.; Garnier, E.; Sastre, B.; Garcia, S.; Mayet, A.; Berdah, S.V. Bioabsorbable Staple-Line Reinforcement for Pancreatectomy in a Porcine Model: A Preliminary Study. *Eur. Surg. Res.* **2012**, *48*, 48–53. [CrossRef]
34. Thomas, P.; Massard, G.; Porte, H.; Doddoli, C.; Ducrocq, X.; Conti, M. A New Bioabsorbable Sleeve for Lung Staple-Line Reinforcement (FOREseal^{TM}): Report of a Three-Center Phase II Clinical Trial. *Eur. J. Cardio-Thorac. Surg.* **2006**, *29*, 880–885. [CrossRef]
35. Schrufer-Poland, T.L.; Ruiz, M.P.; Kassar, S.; Tomassian, C.; Algren, S.D.; Yeast, J.D. Incidence of Wound Complications in Cesarean Deliveries Following Closure with Absorbable Subcuticular Staples versus Conventional Skin Closure Techniques. *Eur. J. Obstet. Gynecol. Reprod. Biol.* **2016**, *206*, 53–56. [CrossRef]
36. Hanna, D.N.; Schlegel, C.; Ghani, M.O.; Hermina, A.; S Mina, A.; McKay, K.; Bailey, C.E.; Magge, D.; Idrees, K. Stapled Full-Thickness Diaphragm Resection: A Novel Approach to Diaphragmatic Resection in Cytoreductive Surgery with Hyperthermic Intraperitoneal Chemotherapy. *J. Am. Coll. Surg.* **2022**, *234*, e1–e6. [CrossRef]
37. Nasir, S. Using Linear Cutting Stapler for Gracillis Muscle Flap Splitting in Facial Reanimation. *Microsurgery* **2018**, *38*, 122–123. [CrossRef]
38. Belfort, M.A.; Shamshiraz, A.A.; Fox, K. Minimizing Blood Loss at Cesarean-Hysterectomy for Placenta Previa Percreta. *Am. J. Obstet. Gynecol.* **2017**, *216*, 78.e1–78.e2. [CrossRef]
39. Kang, S.-K.; Bok, J.S.; Cho, H.J.; Kang, M.-W. Novel Asymmetrical Linear Stapler (NALS) for Pathologic Evaluation of True Resection Margin Tissue. *J. Thorac. Dis.* **2018**, *10*, S1631–S1636. [CrossRef]
40. Salyer, C.; Spuzzillo, A.; Wakefield, D.; Gomaa, D.; Thompson, J.; Goodman, M. Assessment of a Novel Stapler Performance for Laparoscopic Sleeve Gastrectomy. *Surg. Endosc.* **2021**, *35*, 4016–4021. [CrossRef] [PubMed]
41. Nakamura, Y.; Matsumoto, S.; Matsushita, A.; Yoshioka, M.; Shimizu, T.; Yamahatsu, K.; Uchida, E. Pancreaticojejunostomy with Closure of the Pancreatic Stump by Endoscopic Linear Stapler in Laparoscopic Pancreaticoduodenectomy: A Reliable Technique and Benefits for Pancreatic Resection. *Asian J. Endosc. Surg.* **2012**, *5*, 191–194. [CrossRef] [PubMed]
42. Noshiro, H.; Urata, M.; Ikeda, O.; Iwasaki, H.; Nabae, T.; Uchiyama, A.; Nagai, E.; Tanaka, M. Triangulating Stapling Technique for Esophagogastrostomy after Minimally Invasive Esophagectomy. *Surgery* **2013**, *154*, 604–610. [CrossRef]
43. Tsujimoto, H.; Tsuda, H.; Hiraki, S.; Nomura, S.; Ito, N.; Kanematsu, K.; Horiguchi, H.; Aosasa, S.; Yamamoto, J.; Hase, K. In Vivo Evaluation of a Modified Linear Stapling Device Designed to Facilitate Accurate Pathologic Examination of the Surgical Margin. *Gastric Cancer* **2016**, *19*, 666–669. [CrossRef] [PubMed]
44. Chen, Z.; Wang, D.; Zhao, Q.; Yang, P.; Ding, P.; Fan, H.; Dong, T.; Liu, Z.; Yang, X.; Ren, L.; et al. A Case Series of 10 Patients Undergone Linear Cutter/Stapler Guiding Device-Led Overlapped Esophagojejunostomy: A Preliminary Study. *J. Gastrointest. Oncol.* **2023**, *14*, 617–625. [CrossRef] [PubMed]
45. Panther Healthcare Canada, Open Linear Cutter and Cartridges. Available online: https://pantherhealthcarecanada.com/products/open-linear-cutter/ (accessed on 19 July 2023).
46. Medtronic, GIA^{TM} Stapler with DST Series^{TM} Technology. Available online: https://www.medtronic.com/covidien/en-us/support/products/surgical-stapling/dst-series-gia-single-use-reloadable-staplers.html (accessed on 19 July 2023).
47. Ethicon, Linear Cutters. Available online: https://www.ethicon.com/na/epc/code/ntlc55?lang=en-default (accessed on 19 July 2023).
48. Ethicon, Linear Cutters. Available online: https://www.ethicon.com/na/epc/code/ntlc75?lang=en-default (accessed on 19 July 2023).
49. Ethicon, PROXI-MATE® Linear Cutters. Available online: https://www.jnjmedtech.com/en-us/product/ethicon-proximate-linear-cutters (accessed on 19 July 2023).
50. Ethicon, ENDOPATH^{TM} ETS Articulating Linear Cutter. Available online: https://www.jnjmedtech.com/en-us/product/endopath-ets-articulating-linear-cutter (accessed on 19 July 2023).
51. Standard Bariatrics, Standard Titan. Available online: https://standardbariatrics.com/ (accessed on 19 July 2023).
52. Perrone, E.E.; Galganski, L.A.; Tarantal, A.F.; Olstad, K.J.; Treadwell, M.C.; Berman, D.R.; Jarboe, M.D.; Mychaliska, G.B.; Farmer, D.L. Fetal Surgery in the Primate 4.0: A New Technique 30 Years Later. *Fetal Diagn. Ther.* **2021**, *48*, 43–49. [CrossRef]
53. Church, J.T.; McLeod, J.S.; Coughlin, M.A.; Bergin, I.L.; Perkins, E.M.; Hoffman, H.R.; Bilger, M.; Rojas-Peña, A.; Treadwell, M.C.; Berman, D.R.; et al. An Early Investigation into Possible Alternatives to Stapled Hysterotomy in Open Fetal Surgery. *Am. J. Perinatol.* **2019**, *36*, 742–750. [CrossRef]
54. Anthimidis, G. Laparoscopic Excision of a Pedunculated Uterine Leiomyoma in Torsion as a Cause of Acute Abdomen at 10 Weeks of Pregnancy. *Am. J. Case Rep.* **2015**, *16*, 505–508. [CrossRef]
55. Tatekawa, Y.; Kanehiro, H.; Nakajima, Y. Laparoscopic Extirpation of Splenic Hamartoma. *Pediatr. Surg. Int.* **2007**, *23*, 911–914. [CrossRef]

56. Akil, A.; Semik, M.; Freermann, S.; Reichelt, J.; Redwan, B.; Görlich, D.; Fischer, S. Use of a Powered Stapling System for Minimally Invasive Lung Volume Reduction Surgery: Results of a Prospective Double-Blind Single-Center Randomized Trial. *Thorac. Cardiovasc. Surg.* **2019**, *67*, 216–221. [CrossRef] [PubMed]
57. Povoski, S.P. Novel Applications of Endo GIA Linear Staplers during Pancreaticoduodenectomy and Total Pancreatectomy. *Am. J. Surg.* **2001**, *182*, 77–80. [CrossRef] [PubMed]
58. Ji, Y.; Qiu, B.; Gao, S. The Powered Vascular Staple (PVS) versus Conventional Powered Linier Cutter (PLC) for the Application of Bronchial Transection in Thoracoscopic Anatomic Segmentectomy. *J. Thorac. Dis.* **2019**, *11*, 4647–4653. [CrossRef] [PubMed]
59. Miller, D.L.; Roy, S.; Kassis, E.S.; Yadalam, S.; Ramisetti, S.; Johnston, S.S. Impact of Powered and Tissue-Specific Endoscopic Stapling Technology on Clinical and Economic Outcomes of Video-Assisted Thoracic Surgery Lobectomy Procedures: A Retrospective, Observational Study. *Adv. Ther.* **2018**, *35*, 707–723. [CrossRef]
60. Tsunezuka, Y.; Tanaka, N.; Fujimori, H. The Impact of Endoscopic Stapler Selection on Bleeding at the Vascular Stump in Pulmonary Artery Transection. *Med. Devices Evid. Res.* **2020**, *13*, 41–47. [CrossRef]
61. Li, L.; Chen, Y.; Du, J.; Wei, J.; Wang, K.; Zhang, S.; Zhang, N. Pre-Expansion of Posterior Gastric Fascia during Laparoscopic Radical Gastrectomy for Gastric Cancer Prevents Injuries to Upper Short Gastric Vessels. *Transl. Cancer Res.* **2020**, *9*, 358–362. [CrossRef] [PubMed]
62. Charalabopoulos, A.; Davakis, S.; Paraskeva, P.; Machairas, N.; Kapelouzou, A.; Kordzadeh, A.; Sakarellos, P.; Vailas, M.; Baili, E.; Bakoyiannis, C.; et al. Feasibility and Short-Term Outcomes of Three-Dimensional Hand-Sewn Esophago-Jejunal Anastomosis in Completely Laparoscopic Total Gastrectomy for Cancer. *Cancers* **2021**, *13*, 4709. [CrossRef]
63. Panther Healthcare Canada, Endo Linear Cutter and P2G Cartridges. Available online: https://pantherhealthcarecanada.com/products/endo-linear-cutter/ (accessed on 19 July 2023).
64. Medtronic. Available online: https://www.medtronic.com/covidien/en-us/products/surgical-stapling/laparoscopic-staplers.html#endo-gia-ultra-universal-staplers-and-reloads.html (accessed on 19 July 2023).
65. Medtronic, Endo GIA™ Ultra Universal Staplers. Available online: https://www.medtronic.com/covidien/en-us/support/products/surgical-stapling/endo-gia-ultra-universal-staplers-and-reloads.html (accessed on 19 July 2023).
66. Ethicon, ECHELON FLEX™ ENDOPATH®. Available online: https://www.jnjmedtech.com/en-us/product/echelon-flex-endopath-staplers (accessed on 19 July 2023).
67. Guweidhi, A.; Steffen, R.; Metzger, A.; Teuscher, J.; Flückiger, P.; Z'graggen, K. Circular Stapler Introducer. *Dis. Colon. Rectum* **2009**, *52*, 746–748. [CrossRef] [PubMed]
68. Burla, L.; Weibel, P.; Baum, C.; Huber, M.; Gürtler, T.; Weber, M. Linear versus Circular Stapler for Gastrojejunal Anastomosis in Laparoscopic Roux-En-Y Gastric Bypass: An Analysis of 211 Cases. *Surg. Res. Pract.* **2020**, *2020*, 4090797. [CrossRef]
69. Bohdjalian, A.; Langer, F.B.; Kranner, A.; Shakeri-Leidenmühler, S.; Zacherl, J.; Prager, G. Circular- vs. Linear-Stapled Gastrojejunostomy in Laparoscopic Roux-En-Y Gastric Bypass. *Obes. Surg.* **2010**, *20*, 440–446. [CrossRef]
70. Barr, A.C.; Lak, K.L.; Helm, M.C.; Kindel, T.L.; Higgins, R.M.; Gould, J.C. Linear vs. Circular-Stapled Gastrojejunostomy in Roux-En-Y Gastric Bypass. *Surg. Endosc.* **2019**, *33*, 4098–4101. [CrossRef]
71. Giordano, S.; Tolonen, P.; Victorzon, M. Comparison of Linear versus Circular Stapling Techniques in Laparoscopic Gastric Bypass Surgery—A Pilot Study. *Scand. J. Surg.* **2010**, *99*, 127–131. [CrossRef] [PubMed]
72. Ding, W.; Tan, Y.; Xue, W.; Wang, Y.; Xu, X.-Z. Comparison of the Short-Term Outcomes between Delta-Shaped Anastomosis and Conventional Billroth I Anastomosis after Laparoscopic Distal Gastrectomy. *Medicine* **2018**, *97*, e0063. [CrossRef] [PubMed]
73. Milone, M.; Elmore, U.; Manigrasso, M.; Vertaldi, S.; Aprea, G.; Servillo, G.; Parise, P.; De Palma, G.D.; Rosati, R. Circular versus Linear Stapling Oesophagojejunostomy after Laparoscopic Total Gastrectomy. A Systematic Review and Meta-Analysis. *Am. J. Surg.* **2022**, *223*, 884–892. [CrossRef]
74. Raza, S.T. A Circular Surgical Stapler Designed to Anastomose Aorta and Dacron Tube Graft. *AORTA* **2013**, *1*, 71–77. [CrossRef]
75. Giaccaglia, V.; Antonelli, M.S.; Franceschilli, L.; Salvi, P.F.; Gaspari, A.L.; Sileri, P. Different Characteristics of Circular Staplers Make the Difference in Anastomotic Tensile Strength. *J. Mech. Behav. Biomed. Mater.* **2016**, *53*, 295–300. [CrossRef]
76. Edholm, D. Systematic Review and Meta-Analysis of Circular- and Linear-Stapled Gastro-Jejunostomy in Laparoscopic Roux-En-Y Gastric Bypass. *Obes. Surg.* **2019**, *29*, 1946–1953. [CrossRef]
77. Mazaki, J.; Katsumata, K.; Udo, R.; Tago, T.; Kasahara, K.; Kuwabara, H.; Enomoto, M.; Ishizaki, T.; Nagakawa, Y.; Tsuchida, A. Comparison of Pressure Resistance of Double-Rows and Triple-Rows Circular Stapler in Rectal Double Stapling Technique: In Vitro Study. *Medicine* **2022**, *101*, e29600. [CrossRef]
78. Strassner, H.; Caulk, A.; Reher, N.; Petrescu, S.; Vasanji, A. Evaluating Performance of Circular Staplers Us-ing Comparative Test Methods for Evidence-Based Surgery. *Surg. Innov.* **2023**, *30*, 576–585. [CrossRef] [PubMed]
79. Golden Stapler, Single Use Hemorrhoidal Circular Stapler. Available online: http://en.goldenstapler.com/open-surgical-instruments/5.html (accessed on 19 July 2023).
80. Medtronic, EEA™ Circular Stapler with DST™ Series Technology. Available online: https://www.medtronic.com/covidien/en-us/support/products/surgical-stapling/eea-circular-stapler-dst-series-technology.html (accessed on 19 July 2023).
81. Medtronic, Suturatrice Circolare EEA™ Con Tri-Staple™ Technology. Available online: https://www.medtronic.com/covidien/it-it/products/surgical-stapling/eea-circular-stapler.html# (accessed on 19 July 2023).
82. Ethicon, Circular Stapler. Available online: https://www.ethicon.com/emea/epc/code/cdh21a?lang=en-default (accessed on 19 July 2023).

83. Ethicon, Circular Stapler. Available online: https://www.ethicon.com/emea/epc/code/ecs21a?lang=en-default (accessed on 19 July 2023).
84. Ethicon, Circular Stapler. Available online: https://www.ethicon.com/na/epc/code/ecs21b?lang=en-default (accessed on 19 July 2023).
85. Sun, D.; Zhang, R.; Wei, M.; Liu, P.; Zhong, X.; Liang, Y.; Chen, Y.; Huang, Y.; Yu, W. Comparison Between Linear Stapler and Circular Stapler After Laparoscopic-Assisted Distal Gastrectomy in Patients With Gastric Cancer. *Front. Surg.* **2022**, *9*, 858236. [CrossRef]
86. Brisinda, G.; Chiarello, M.M.; Pepe, G.; Cariati, M.; Fico, V.; Mirco, P.; Bianchi, V. Anastomotic Leakage in Rectal Cancer Surgery: Retrospective Analysis of Risk Factors. *World J. Clin. Cases* **2022**, *10*, 13321–13336. [CrossRef]
87. Sinolinks, Disposable Circular Stapler. Available online: https://sinolinks.com.cn/stapler-en/102.html (accessed on 19 July 2023).
88. Covidien, EEA™ Hemorrhoid and Prolapse Stapler Set. Available online: https://www.medtronic.com/covidien/en-us/support/products/surgical-stapling/eea-hemorrhoid-and-prolapse-stapler-set.html (accessed on 19 July 2023).
89. Ethicon, PROXIMATE™ PPH Hemorrhoidal Circular Stapler Set. Available online: https://www.jnjmedtech.com/en-us/product/proximate-pph-hemorrhoidal-circular-stapler-set (accessed on 19 July 2023).
90. Echelon Flex Powered Vascular Stapler. Ethicon. Cincinnati. Available online: https://www.jnjmedtech.com/en-us/product/echelon-flex-powered-vascular-stapler (accessed on 19 July 2023).
91. Echelon Circular Powered Vascular Stapler. Ethicon. Cincinnati. Available online: https://www.jnjmedtech.com/en-us/product/echelon-circular-powered-stapler (accessed on 19 July 2023).
92. Sonoda, T.; Verdeja, J.C.; Rivadeneira, D.E. Stapler Access and Visibility in the Deep Pelvis: A Comparative Human Cadaver Study between a Computerized Right Angle Linear Cutter versus a Curved Cutting Stapler. *Ann. Surg. Innov. Res.* **2011**, *5*, 7. [CrossRef]
93. Shimada, Y.; Maehara, S.; Osawa, J.; Hagiwara, M.; Ohira, T.; Ikeda, N. Powered articulation by the SigniaTM stapling system for stapling position adjustments: Optimizing safe surgical margins in thoracoscopic sublobar resection. *Surg. Today* **2021**, *51*, 447–451. [CrossRef] [PubMed]
94. Roy, S.; Yoo, A.; Yadalam, S.; Fegelman, E.J.; Kalsekar, I.; Johnston, S.S. Comparison of Economic and Clinical Outcomes between Patients Undergoing Laparoscopic Bariatric Surgery with Powered versus Manual Endoscopic Surgical Staplers. *J. Med. Econ.* **2017**, *20*, 423–433. [CrossRef]
95. Park, S.Y.; Kim, D.J.; Mo Nam, C.; Park, G.; Byun, G.; Park, H.; Choi, J.H. Clinical and Economic Benefits Associated with the Use of Powered and Tissue-Specific Endoscopic Staplers among the Patients Undergoing Thoracoscopic Lobectomy for Lung Cancer. *J. Med. Econ.* **2019**, *22*, 1274–1280. [CrossRef]
96. Eckert, C.E.; Harris, J.L.; Wong, J.; Thompson, S.; Kassis, E.S.; Tsuboi, M.; Ott, H.C.; Force, S. Preclinical Quantification of Air Leaks in a Physiologic Lung Model: Effects of Ventilation Modality and Staple Design. *Med. Devices Evid. Res.* **2018**, *11*, 433–442. [CrossRef] [PubMed]
97. Rawlins, L.; Johnson, B.H.; Johnston, S.S.; Elangovanraaj, N.; Bhandari, M.; Cohen, R.V.; Rheinwalt, K.P.; Fryrear, R.; Roy, S. Comparative Effectiveness Assessment of Two Powered Surgical Stapling Platforms in Laparoscopic Sleeve Gastrectomy: A Retrospective Matched Study. *Med. Devices Evid. Res.* **2020**, *13*, 195–204. [CrossRef] [PubMed]
98. Pla-Martí, V.; Martín-Arévalo, J.; Moro-Valdezate, D.; García-Botello, S.; Mora-Oliver, I.; Gadea-Mateo, R.; Cozar-Lozano, C.; Espí-Macías, A. Impact of the Novel Powered Circular Stapler on Risk of Anastomotic Leakage in Colorectal Anastomosis: A Propensity Score-Matched Study. *Tech. Coloproctol.* **2021**, *25*, 279–284. [CrossRef]
99. Herzig, D.O.; Ogilvie, J.W.; Chudzinski, A.; Ferrara, A.; Ashraf, S.Q.; Jimenez-Rodriguez, R.M.; Van der Speeten, K.; Kinross, J.; Schimmelpenning, H.; Sagar, P.M.; et al. Assessment of a Circular Powered Stapler for Creation of Anastomosis in Left-Sided Colorectal Surgery: A Prospective Cohort Study. *Int. J. Surg.* **2020**, *84*, 140–146. [CrossRef]
100. Pollack, E.; Johnston, S.; Petraiuolo, W.J.; Roy, S.; Galvain, T. Economic Analysis of Leak Complications in Anastomoses Performed with Powered versus Manual Circular Staplers in Left-Sided Colorectal Resections: A US-Based Cost Analysis. *Clin. Outcomes Res.* **2021**, *13*, 531–540. [CrossRef]
101. Sylla, P.; Sagar, P.; Johnston, S.S.; Dwarakanathan, H.R.; Waggoner, J.R.; Schwiers, M.; Roy, S. Outcomes Associated with the Use of a New Powered Circular Stapler for Left-Sided Colorectal Reconstructions: A Propensity Score Matching-Adjusted Indirect Comparison with Manual Circular Staplers. *Surg. Endosc.* **2022**, *36*, 2541–2553. [CrossRef]
102. Medtronic Covidien Signia Stapling Platform. Covidien Japan, Tokyo. Available online: https://www.medtronic.com/covidien/it-it/products/surgical-stapling/signia-stapling-system.html# (accessed on 19 July 2023).

applied sciences

MDPI

Article

Biopringing of a Biomimetic Microenvironment for a Retinal Regenerative Approach

Beatrice Belgio, Sara Mantero, Filippo Iervolino, Federica Potere, Marinella Levi and Federica Boschetti *

Department of Chemistry, Materials and Chemical Engineering "Giulio Natta", Politecnico di Milano, 20133 Milan, Italy; beatrice.belgio@polimi.it (B.B.); sara.mantero@polimi.it (S.M.); filippo.iervolino@polimi.it (F.I.); federica.potere@polimi.it (F.P.); marinella.levi@polimi.it (M.L.)

* Correspondence: federica.boschetti@polimi.it

Abstract: There is an ongoing effort to advance methodologies for culturing functional photoreceptors in vitro for retinal regenerative strategies. To support the formation of functional photoreceptors, a scaffold should replicate the native environment. The aim of this study was to optimize a sodium alginate–gelatin (SA-G) bioink to mimic the retinal properties while ensuring the printing of constructs with high shape fidelity. The optimized bioink was thoroughly characterized in terms of its physical, mechanical, and rheological properties, printability assessment, and preliminary biocompatibility. The material showed a constant degradation rate, which is crucial for effective tissue regeneration as it provides support for cell differentiation and polarization while gradually degrading to allow cell proliferation and matrix deposition. The optimized bioink displayed stiffness comparable to the native photoreceptor layer, potentially providing appropriate mechanical cues for photoreceptor maturation. Additionally, it exhibited shear-thinning behavior, the presence of yield stress, and fast recovery kinetics, which are essential for successful extrusion. The high shape fidelity of 3D-printed constructs suggested the feasibility of printing complex patterns to drive photoreceptor polarization. The preliminary cell results demonstrated homogeneous cell distribution and sustained cell viability over time. Overall, these findings indicate that the optimized bioink can provide the mechanical and topographical cues necessary for cultivating photoreceptors in vitro for retinal regeneration.

Keywords: direct ink writing; sodium alginate; gelatin; retinal regeneration; bioink; rheological characterization

1. Introduction

Many pathologies affecting retinal tissue lead to irreversible vision impairments due to the loss of photoreceptors. Regenerative medicine represents a viable strategy for treating these conditions by replacing dead photoreceptors with healthy ones. Research advancements in cell therapies have shown promise in restoring retinal cells and retinal function [1]. For instance, the use of induced pluripotent stem cells (iPSCs) to generate a functional retinal pigment epithelium has demonstrated potential in clinical trials [2]. Yet, significant hurdles, including the long-term survival and integration of transplanted cells, remain [3]. Moreover, photoreceptors are difficult to cultivate and maintain in vitro [4]. To tackle these issues, cells can be transplanted using a scaffold, allowing for the delivery of an organized and functional cell layer that integrates better with the host tissue compared with cell injection. Yet, few studies on the fabrication of photoreceptor scaffolds have been reported in the literature. Most of these studies focused on creating scaffolds with an architecture that guides cell differentiation and orientation, which is crucial for fulfilling the light-sensing function [5]. Steedman et al. proved that the presence of microtopography positively affected cell attachment and induced the differentiation of retinal progenitor cells [6]. Similarly, Jung et al. showed that a 3D micropatterned scaffold can guide the attachment and the differentiation of pluripotent stem cells into photoreceptors [7]. Besides topographical cues, scaffold stiffness has also been identified as a key parameter for driving cells to differentiate into photoreceptors and for maintaining lineage specification [8,9]. Therefore, selecting suitable substrate materials and fabrication techniques is crucial for recapitulating both the architecture and properties of the in vivo photoreceptor environment [8]. Current substrates fail to mimic both the architecture and the stiffness of the native milieu, thus hindering progress in retinal regenerative approaches [7,10]. Hence, recreating the environment of photoreceptors in vitro still represents an engineering challenge.

Recently, 3D bioprinting has been suggested as a promising technique to fabricate scaffolds for photoreceptors, as it allows 3D porous architectures with precise control over pore size, geometry, and cell spatial distribution to be obtained without significantly affecting cell viability [5,11–13]. Different bioprinting technologies have been developed and employed according to the application [14]. For instance, Masaeli et al. demonstrated the feasibility of fabricating a 3D in vitro retina model using an inkjet-based bioprinter [12]. However, with this approach, cells had to be printed in culture medium to avoid clogging issues. So, the native stiffness could not be matched. Among the bioprinting approaches, pneumatic-driven extrusion-based bioprinting is one of the most common due to its versatility in printing a wide range of bioinks and its ease of use [15]. Additionally, it enables the production of clinically relevant constructs, in terms of size, within a reasonable time. Bioinks employed in extrusion bioprinting must display shear-thinning behavior to enable successful extrusion and printing while preserving the structure given by the printing process [16]. Shi et al. used an alginate–pluronic bioink that included a retinoblastoma-derived cell line to replicate the photoreceptor layer in a 3D in vitro model [17]. They evaluated the impact of two different printing patterns on cell morphology; however, the stiffness of the bioink was not investigated. Hence, to date, a bioink capable of meeting the requirements for extrusion-based bioprinting and simultaneously accurately replicating retinal properties has yet to be developed [11,18].

Sodium alginate–gelatin (SA-G) blends have been successfully applied over the years in extrusion-based 3D bioprinting for engineering different tissues, such as liver and neural tissue [19–21]. These blends have shown a good printability window, a rapid crosslinking rate, printing accuracy, and biocompatibility [22,23]. Moreover, they can be obtained via an easy and cost-effective process. We believe that an SA-G bioink holds great potential for bioprinting a functional photoreceptor layer, as gelatin provides cell attachment sites, while sodium alginate exhibits a suitable stiffness for the development of photoreceptors. Hunt et al. found that a sodium alginate-based hydrogel functionalized with arginine–glycine–aspartate groups promoted photoreceptor cell differentiation in comparison with hydrogels based on hyaluronic acid and pure gelatin due to its stiffness [8].

The present research aims to introduce a novel platform that is potentially able to guide and sustain functional photoreceptors in vitro through both topographical and mechanical cues. To this end, we optimized a bioink composed of sodium alginate and gelatin to replicate retinal native properties, such as stiffness, while ensuring successful extrusion and high shape fidelity in printed constructs. To the best of our knowledge, no bioink with these features has been developed previously. To investigate the properties of the optimized SA-G bioink, we performed a comprehensive characterization based on well-established testing protocols as well as protocols developed specifically for this study.

2. Materials and Methods

A schematic representation of the research methodology is reported in Figure 1.

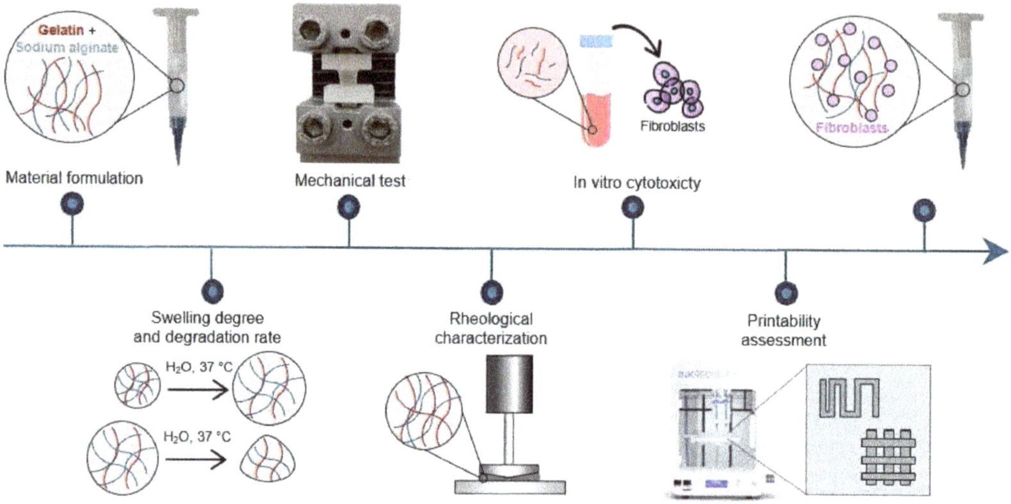

Figure 1. Graphic representation of the research methodology.

2.1. Material Preparation

To prepare the SA-G bioink, low-viscosity sodium alginate (154725, Lot Number: SR01864) and gelatin (bovine skin type B, G9391, Lot Number: SLCF9893) powders were purchased from MP Biomedicals (Santa Ana, CA, USA) and Sigma-Aldrich (St. Louis, MO, USA), respectively.

The hydrogel precursor solution was prepared as follows. First, the powders were separately sterilized in 99.99% ethanol (Sigma-Aldrich, St. Louis, MO, USA) for 15 min. The sterilization process was repeated three times. After the complete evaporation of ethanol, 0.4 g of sodium alginate powders were dissolved in 10 mL of sterile phosphate-buffered saline (PBS, 1X, *w/o* Calcium, *w/o* Magnesium, pH = 7.4, Gibco, Thermo Fisher Scientific, Waltham, MA, USA) overnight under vigorous magnetic stirring. Similarly, 1 g of gelatin was dissolved in 10 mL of sterile PBS at 37 °C until total solubilization. The two resulting solutions were mixed together in a volume ratio of 1:1. Hence, the final alginate and gelatin concentrations in the precursor were 2% and 5% *wt./v*. These concentrations were selected after an optimization process based on preliminary extrudability and printability tests. The precursor solution was stored at 4 °C and used within one week.

The crosslinking solution was prepared by dissolving 2% *wt./v* of calcium chloride ($CaCl_2$, Sigma-Aldrich, St. Louis, MO, USA) into sterile distilled water and stored at 4 °C until use. The $CaCl_2$ concentration and the crosslinking duration were optimized based on previous studies [24]. The use of $CaCl_2$ is widely reported in the literature to crosslink sodium alginate [25]. The crosslinking mechanism is based on ionic bonds between calcium ions and carboxylate ions present in the alginate. Briefly, sodium alginate is a linear polysaccharide containing carboxylate ions, each bonded with a sodium ion. When alginate is immersed in a $CaCl_2$ solution, the dissociated calcium ions replace the sodium ions in the alginate, ionically bonding with the carboxylate ions. Since each calcium ion can bond with two carboxylate ions, the process results in the crosslinking of the polymer chains and, thus, in the formation of a hydrogel [25].

2.2. Swelling Degree and Degradation Rate

To evaluate the hydrogel swelling and degradation properties, the SA-G precursor solution was transferred into a Petri dish and crosslinked at 4 °C for 10 min. Hydrogel cylindrical samples (n = 3) of 7 mm diameter and 3 mm thickness were obtained by using a biopsy punch. The samples were then crosslinked by submerging them in the $CaCl_2$ crosslinking solution for 10 min at room temperature.

To evaluate the swelling degrees and degradation rates, the samples were placed in a 6-well plate, immersed in complete culture medium composed of high glucose Dulbecco's Modified Eagle's Medium (DMEM, Gibco™, Thermo Fisher Scientific, Waltham, MA, USA) supplemented with 10% fetal bovine serum albumin (FBS, American Type Culture Collection, Manassas, VA, USA) and 1% penicillin/streptomycin antibiotics (Gibco™, Thermo Fisher Scientific, Waltham, MA, USA), and incubated at 37 °C for the following time points: 15 min, 30 min, 1 h, 2 h, 3 h, 4 h, 5 h, and 6 h (same set of samples for each time point) [26]. After specific incubation periods, the samples were weighed to determine the swelling degree. The swelling degree at each time point was computed according to Equation (1):

$$\text{Swelling Degree}[\%] = \frac{W_t - W_0}{W_0} \times 100 \tag{1}$$

where W_t is the sample weight at each time point, and W_0 is the starting sample weight before adding the solution.

The same set of samples was employed to study the degradation rate. In more detail, after reaching the swelling equilibrium, samples were incubated at 37 °C for 1, 3, 5, 7, 10, 14, and 21 days. At each time point, the samples were weighed, and the degradation rate was calculated by quantifying the weight decrease as in Equation (2):

$$\text{Degradation rate}[\%] = (100 - \frac{W_s - W_t}{W_s}) \times 100 \tag{2}$$

where W_t is the sample weight at each time point, and W_s is the sample weight reached at swelling equilibrium.

2.3. Mechanical Tests

Hydrogel mechanical properties were measured by performing uniaxial tensile tests through an ad hoc setup. This includes two poly(lactic acid) (PLA, K Kentstrapper, Florence, Italy) plates fabricated using fused-deposition modeling (Verve, Kentstrapper, Florence, Italy). Each plate consists of a base and a protrusion with dimensions of (30 × 30 × 10) mm and (10 × 10 × 20) mm, respectively (Figure 2a). The protrusions were designed to be inserted and secured into the jaws of a testing machine (Synergie 200, MTS Systems, Eden Prairie, MN, USA) by tightening, thus preventing any slipping of the plates. Hydrogel samples (n = 5) obtained as described in Section 2.2 were tightly glued to the bases of the plates by using a biocompatible tissue adhesive (3M™ Vetbond™, Thermo Fisher Scientific, Waltham, MA, USA). Subsequently, the plates along with the sample were mounted onto the testing machine, which was equipped with a 100N loading cell (Figure 2b). Samples were pulled until failure at a displacement rate of 0.1 mm/s at room temperature (25 °C). The displacement rate was selected based on previous tensile tests on retinal samples [27].

For the analyses, force (N)−elongation (mm) data were elaborated to obtain the stress [kPa]−strain [mm/mm] relationship for each sample. The stress and the strain were computed as indicated in Equations (3) and (4), respectively. The elastic modulus (E) was extracted from the stress–strain plots as the slope of the initial linear region of the stress–strain curve.

$$\sigma = \frac{F}{A} \tag{3}$$

$$\varepsilon = \frac{\Delta L}{L_0} \tag{4}$$

where σ and ε are the stress and the deformation, respectively. F is the force applied to the original cross section A, and ΔL is the change in length with respect to the original length L_0.

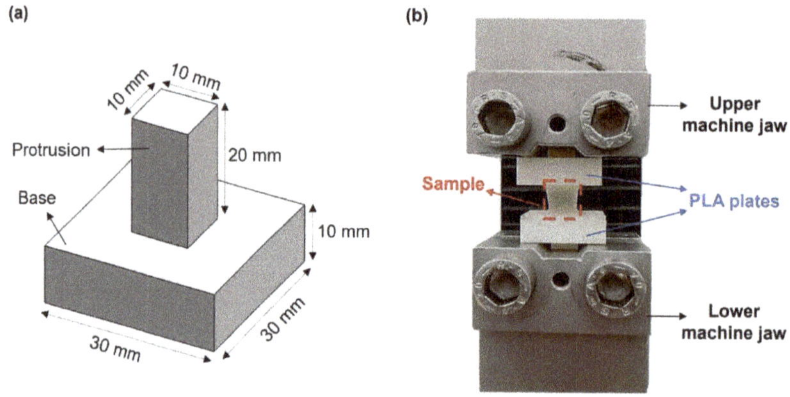

Figure 2. (**a**) CAD model of the poly(lactic acid) (PLA) plates designed to perform uniaxial tensile tests; (**b**) Experimental setup used for the uniaxial tensile tests.

2.4. Rheological Characterization

Rheological analyses were carried out at 25 °C using a rotational rheometer (Discovery HR2, TA Instruments, New Castle, DE, USA) equipped with a cone-plate geometry (diameter: 20 mm). The testing temperature was selected based on the printing temperature. Prior to all tests, a conditioning step, consisting of a rotation of the tool at 10 s^{-1} for 30 s followed by a soak period of 30 s, was included.

To evaluate the rheological properties of the SA-G bioink, two specific tests were performed: a flow curve and a strain sweep. In the former, the viscosity and the shear stress exerted on the material in response to the applied shear rates were measured in a continuous flow experiment. Specifically, the shear rate was progressively increased in a logarithmic manner from 0.001 to 10,000 s^{-1}. Ten points per decade were registered, and the test time was 180 s., i.e., 30 s per decade. In the latter, material responses, namely the storage modulus (G′) and the loss modulus (G″), were measured by applying oscillatory strains in a logarithmic manner from 0.01% to 1000% at a frequency of 1 Hz. Ten points per decade were recorded. The testing parameters for the flow curve and strain sweep were determined by conducting preliminary tests. To assess the impact of cells on the bioink rheological properties, both tests were repeated on the SA-G bioink loaded with L929 cells at different cell densities (5×10^6 cells/mL and 10×10^6 cells/mL) [16]. The cell encapsulation procedure will be illustrated in Section 2.7.

To investigate the material recovery ability after shear application and removal, a strain recovery test and an elastic recovery test were carried out. In the strain recovery test, the viscosity was measured during the following consecutive steps that mimic the printing process: (i) pre-printing phase (shear rate of 0.05 s^{-1} for 300 s); (ii) printing (sudden increase in shear rate to 811 s^{-1} for 60 s); (iii) post-printing phase (shear rate of 0.05 s^{-1} for 300 s). Thirty points per step were recorded. A 0.05 s^{-1} shear rate was chosen to simulate material resting conditions, whereas 811 s^{-1} was found to be the maximum shear rate value applied to the material during extrusion. It was computed by using Equation (5), which was derived from the Hagen–Poiseuille law with Rabinowitsch correction [28].

$$\text{Shear rate} = \frac{4Q}{\pi R^3} \times \frac{3n+1}{4n} \tag{5}$$

where Q (mm^3s^{-1}) is the flow rate, R (mm) is the inner radius at the outlet of the nozzle used in the bioprinting process and (n − 1) is the slope of the viscosity versus shear rate graph on a log–log plot obtained from the flow curve. To assess the material elastic recovery after the printing process, oscillatory time sweeps made at a frequency of 1 Hz were performed with alternating high/low shear stresses, i.e., 7×10^{-4} and 10^{-6} MPa, respectively. The maximum shear stress value set in this test was the one corresponding to a shear rate of 811 s^{-1}. The material was allowed to recover for 2 min to simulate the non-printing phases. All rheological measurements were repeated three times.

2.5. In Vitro Cytotoxicity

Cylindrical hydrogel samples (n = 3), prepared as described in Section 2.2, were placed in 6-well plates and incubated in complete culture medium at 37 °C and 5% CO_2. The medium was collected at different incubation times (1, 5, and 7 days). Meanwhile, L929, a murine fibroblast cell line (Catalog No. CCL-1^{TM}, American Type Culture Collection, Manassas, VA, USA), was routinely cultured in complete culture medium at 37 °C and 5% CO_2, with medium replacement every 2 days. The cells were cultured for four sequential passages until they reached the desired density. L929 cells were then seeded in 12-well plates at a density of 1×10^5 cell/well and grown until they reached 70% confluency. Subsequently, these cells were cultured for 24 h with the medium that had been incubated with the hydrogel samples. As a control, cells were cultured with fresh complete culture medium. All experiments were performed in replicates of three (n = 3). After 24 h, cell viability was assessed using the alamarBlue^{TM} assay (Invitrogen^{TM}, Carlsbad, CA, USA) according to the manufacturer's instruction. Alamar blue is an oxidation–reduction indicator that changes color upon reduction by living cells. To perform the assay, the alamar blue stock solution was diluted to 1:10 with complete culture medium. The resulting 10% alamar blue solution was incubated with the cells for 3 h at 37 °C. After incubation, a volume of 100 µL/well was transferred to a 96-well plate for absorbance reading. The absorbance at 570 nm and 600 nm was measured using a microplate reader (infinite 200Pro, Tecan Group Ltd, Männedorf, Switzerland), and the percentage viability was computed as the percent difference in reduction between treated and control cells.

2.6. Printability Assessment

Cellink INKREDIBLE+ (Cellink AB, Gothenburg, Sweden) was used to print the SA-G bioink. Such a 3D bioprinter is based on pneumatic extrusion technology and is equipped with dual printheads featuring a heating system and built-in photocuring modules. Moreover, it includes patented clean-chamber technology that efficiently filters air through an H13 HEPA filter, thus providing a clean printing environment. All structures introduced in this section were designed with Solidworks 2020 (Dassault Systems Solid WorksCorp, Waltham, MA, USA), sliced with PrusaSlicer 2.3.3. (Prusa Research, Prague, Czech Republic), and printed at 25 °C. Two conical nozzles with inner diameters of 22G (0.41 mm) and 25G (0.25 mm) were used. Prior to printing, the hydrogel precursor solution stored at 4 °C was incubated at 37 °C for 15 min, thus allowing the solution to be transferred to a printing cartridge. The printing cartridge was left at 25 °C for 40 min before printing to enable partial gelatin crosslinking. After printing, all structures were physically crosslinked at 4 °C for 5 min followed by chemical crosslinking in a 2% *wt./v* $CaCl_2$ bath for 10 min.

Printability was assessed by evaluating the filament formation and printing accuracy in terms of filament merging and shape fidelity in multi-layered structures [16]. For the first test mentioned, four lines were printed at different increasing velocities, i.e., 5, 10, 15, and 20 mm/s, by applying a pressure of 25 kPa. The test was repeated using a printing pressure of 30 kPa. The line widths were measured using ImageJ software (NIH, Stapleton, NY, USA, ver. 1.54i) after imaging the lines with an optical microscope (Eclipse Ti2, Nikon, Tokyo, Japan). Based on the results of this test, optimized printing parameters were chosen for further analyses. The printing accuracy on the x–y plane was determined through the filament fusion test [29]. For this test, a precise pattern was designed that involved parallel lines spaced at different distances (from 0.5 mm to 3 mm), with a progressive 0.5 mm increase for each subsequent line. Ultimately, circular discs with a diameter of 30 mm and a thickness of 0.4 mm were designed and printed with four layers to evaluate the shape fidelity of multilayered constructs. Specifically, the dimensions of the printed structures were measured using ImageJ software and compared with the original dimensions set in the digital model of the structure. Moreover, optical images taken by the Eclipse Ti2 microscope were analyzed to investigate the pore geometry of the printed discs. Subsequently, the printability index (Pr) was computed using Equation (6) [16]:

$$Pr = \frac{L^2}{16 \times A} \quad (6)$$

where L and A are the pore perimeter and area, respectively.

2.7. Bioprinting

Before bioprinting, L929 cells routinely cultured in complete culture medium were centrifuged and resuspended in fresh culture medium at a density of 5×10^6 cells/mL. The cell suspension was mixed with the SA-G precursor solution previously heated at 37 °C for 15 min by means of a luer-lock connector to homogeneously encapsulate cells within the solution. The SA-G bioink containing cells was transferred into a printing cartridge equipped with a 22G (0.41 mm) conical nozzle and left at 25 °C for 40 min before printing. Four-layered grids with dimensions of 10 mm × 10 mm × 1 mm were then printed using the printing parameters established with the printability tests (Section 2.6). The printed structures were crosslinked as indicated in Section 2.1 and cultured at 37 °C and 5% CO_2. Cell viability was evaluated with the LIVE/DEAD Cell Viability Assay (Invitrogen, Carlsbad, CA, USA) following manufacturer instructions. Fluorescent images were acquired with an optical microscope (Eclipse Ti2, Nikon, Tokyo, Japan) 1 and 5 days after printing.

2.8. Statistical Analyses

Statistical analyses were performed with IBM SPSS software (ver. 29). Results are presented as mean ± SD. To evaluate the effect of cell density on the SA-G viscosity, the Friedman test was used with a 95% confidence interval, whereas for the cytotoxicity analysis of degradation products, two-way ANOVA was employed with a 95% confidence interval. Significant differences were reported for p-values lower than 0.05, with * indicating a p-value < 0.05.

3. Results and Discussion

3.1. Swelling Degree and Degradation Rate

Hydrogel swelling ability is crucial for the penetration of nutrients and for the transport of metabolic products, thus promoting cell viability and proliferation within the hydrogel [30]. Moreover, when developing a hydrogel as a bioink, its swelling behavior needs to be evaluated and optimized to preserve the shape fidelity of the printed structures once immersed in culture medium [16]. Figure 3a displays the average swelling degree of the SA-G hydrogel samples. Within the first hour, the average degree of swelling increased to approximately 55%. Then, the hydrogel samples continued swelling at a slower rate with an average rise of 20% until the sixth hour, when they reached the swelling plateau. The maximum swelling degree value was 80%, on average. Such a value allows for the uptake and transport of nutrients dissolved in the cell culture medium, thus favoring cell viability and proliferation, while retaining shape fidelity of the printed constructs. Indeed, based on the findings of Li et al., bioinks characterized by a swelling degree of nearly 100% did not show significant spreading of filaments and pores in the printed constructs [31].

As for the degradation rate, the material should progressively degrade to allow cells to proliferate and deposit new extracellular matrix while preserving the initial structure of the printed constructs [32]. The SA-G hydrogel samples exhibited a constant weight decrease over 21 days (Figure 3b). On day 21, the average weight decrease was 16%, although few samples were completely degraded.

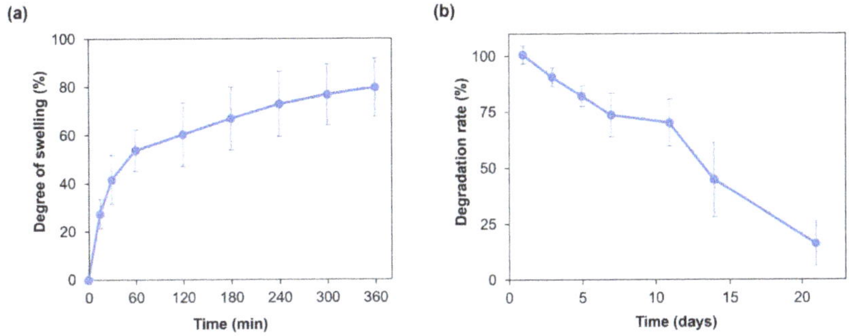

Figure 3. (**a**) Swelling degree (%) of the SA-G hydrogels over time; (**b**) Degradation rate (%) of the SA-G hydrogels over time. Error bars represent the standard deviation. SA-G: sodium alginate–gelatin.

3.2. Mechanical Testing

The mechanical properties of the SA-G hydrogel samples were studied through uniaxial tensile testing, as knowledge about retinal mechanical properties is mainly based on uniaxial tensile tests [33]. Since the conventional grips used for uniaxial tensile testing exert localized pressure on soft materials, leading to significant local deformations and fractures, a custom setup able to constrain the sample without causing clamping-induced fractures was designed and developed. The sample fracture site at the end of mechanical testing proved the effectiveness of the new setup (Figure 4a). Figure 4b shows a typical tensile stress−strain curve of the tested SA-G hydrogels. The curve presents a linear elastic region followed by failure, consistent with the literature [34]. As observed in Figure 4b, the fracture was not instantaneous. The stress dropped gradually, indicating a progressive failure of the hydrogel. This behavior may be due to the presence of regions with slightly different crosslinking degrees, although further studies are needed to confirm this hypothesis. The average elastic modulus (E), extrapolated from the stress–strain graph, was 37.5 $\pm$ 0.36 kPa. According to our previous results, the average E of porcine retinal tissue at room temperature is lower (13.4 $\pm$ 0.0067 kPa) [27]. However, the test was performed ex vivo on the entire retinal tissue. Qu et al. exploited a phase-resolved acoustic radiation force optical coherence elastography method to determine the elastic properties of different porcine retinal layers in vivo. They found that the stiffness increased from the ganglion layer to the photoreceptors, with the maximum value of 25.9 $\pm$ 7.36 kPa in the photoreceptor layer [35]. Hence, the stiffness of the SA-G bioink is akin to the stiffness of the native photoreceptor layer. Thus, it can provide in vivo-like mechanical cues, which are essential for promoting cell differentiation into photoreceptor phenotypes [8].

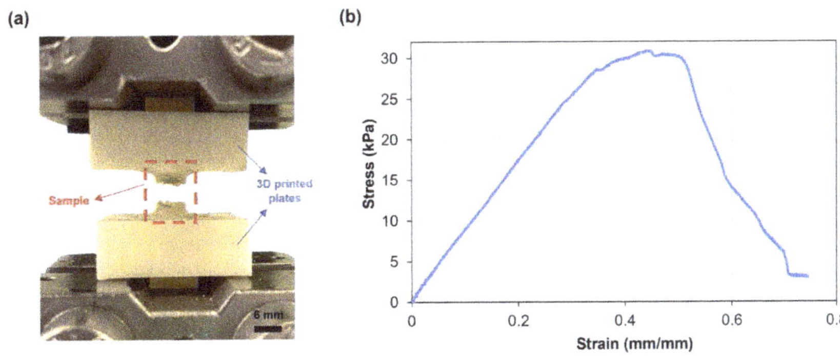

Figure 4. (**a**) SA-G hydrogel samples after testing. Scale bar = 6 mm; (**b**) Representative stress–strain curve of hydrogel samples when pulled until failure. SA-G: sodium alginate–gelatin.

3.3. Rheological Characterization

The study of bioink rheological properties is a prerequisite for a successful extrusion-based process. Indeed, it enables a preliminary evaluation of the bioink extrudability and printability [16]. Therefore, SA-G bioink rheological behavior, including its viscosity, viscoelastic properties, and shear and elastic recovery, was deeply investigated.

The flow curves performed on the SA-G precursor solutions with and without cells are displayed in Figure 5a. From these curves, the presence of a pseudoplastic (or shear thinning) behavior, i.e., a decrease in the material viscosity when applying a shear stress, can be evaluated. The pseudoplastic behavior is a requirement that a bioink must fulfill to be suitable for extrusion-based 3D bioprinting [16]. Specifically, shear-thinning behavior is necessary to extrude with ease a material that has a relatively high zero-shear-rate viscosity (i.e., the viscosity of the material when no shear rate is applied) through a narrow nozzle. Indeed, during the extrusion, a high shear rate is applied to the material, thus reducing its viscosity and allowing for a more uniform filament deposition. As for the SA-G solution without cells, a zero-shear-rate viscosity plateau of 1000 Pa·s was encountered, followed by shear-thinning behavior starting from 0.1 s^{-1}. The presence of cells did not significantly impact the zero-shear-rate viscosity value nor the pseudoplastic behavior, in agreement with the findings of Jain et al. [36]. Another desirable rheological feature of a bioink is represented by the presence of a yield stress, which is the stress value at which the material starts flowing and, therefore, an indication of the material's ability to maintain its own shape after the extrusion process. A high yield stress will prevent the printed structure from collapsing before cross-linking occurs. Such a characteristic was evaluated by plotting the viscoelastic properties, namely the storage modulus (G′) and loss modulus (G″), of the SA-G solutions (Figure 5b). For all solutions, at low strain values, G′ was higher than G″, thus implying that the materials have solid-like behavior, whereas starting from a strain of approximately 800%, G″ was higher than G′, resulting in liquid-like behavior of the materials. The point at which G′ and G″ have the same value is called the crossover point, and it indicates the presence of a yield stress. Indeed, after the crossover point, the material can flow, thus enabling extrusion during the 3D printing process. All inks exhibited a crossover, thus showing the presence of a clear yield stress and, hence, their extrudability. No significant differences in G′ and G″ values were encountered between cell-laden bioinks and the bioink without cells.

Recovery tests were performed to evaluate the material's ability to regain its original properties after printing. Since rheological properties were not affected by the presence of cells at any cell density, these tests were carried out on the SA-G bioink without cells. The results of the strain recovery test are presented in Figure 5c. The viscosity at resting conditions was about 1000 Pa·s; then, when a high value of shear rate was applied, the viscosity value dramatically dropped. These results agreed with the previous findings regarding zero-shear-rate viscosity and pseudoplastic behavior. Once the high shear was removed, the viscosity returned to its initial resting value after a small transition region. Such an outcome proved that the SA-G material has fast recovery kinetics, which is essential to ensure high shape fidelity. The good material recovery ability was confirmed by the elastic recovery test, where rapid transitions from solid-like to liquid-like behavior were found when alternating stress values were applied (Figure 5d).

Taken together, these findings demonstrate that the SA-G bioink is a well-suited material for extrusion-based 3D bioprinting [37].

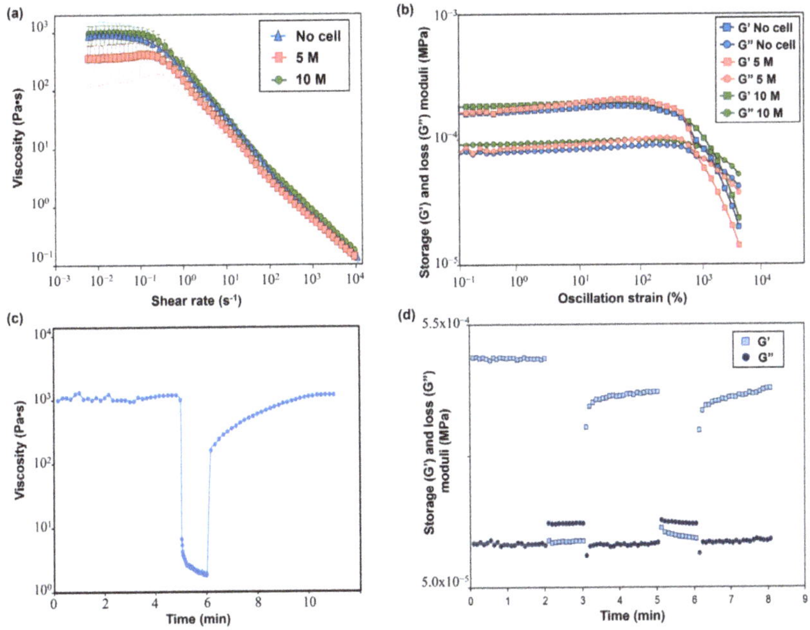

Figure 5. (**a**) Flow curves showing the viscosity–shear rate behavior of SA-G bioinks with and without cells. Error bars represent the standard deviation. Analyses were performed on SA-G bioinks with no cells, with cells included at a density of 5 M cells/mL, and with cells included at a density of 10 M cells/mL; (**b**) Results of strain sweep tests displaying the storage (G′) and loss (G″) moduli against the oscillation strain. G′ and G″ are represented by square markers and circle markers, respectively. Analyses were performed on SA-G bioinks with no cells, with cells included at a density of 5 M cells/mL, and with cells included at a density of 10 M cells/mL. (**c**) Strain recovery test results; (**d**) Elastic recovery test results. G′ and G″ are represented by square markers and circle markers, respectively. SA-G: sodium alginate–gelatin.

3.4. In Vitro Cytotoxicity

As for the cytotoxicity of the products released during material degradation, cell viability was analyzed after culturing cells in contact with eluates of the SA-G hydrogel samples for 1, 5, and 7 days. The time points were chosen according to the results of the degradation test (Section 3.1). Cell viability was not significantly affected by the presence of the degradation products (Figure 6). Indeed, at every time point, the cell viability was found to be higher than 90%. Additionally, no statistical difference was found between the viability of the controls and of the cells in contact with SA-G eluates at any time point (p-value = 0.07). This result suggests that the SA-G bioink is non-cytotoxic and, hence, can be loaded with cells for 3D bioprinting applications.

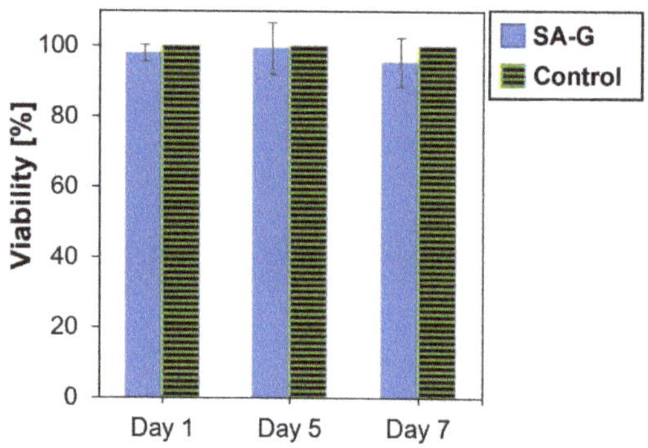

Figure 6. Viability (%) of cells determined by the alamarBlue assay after 1, 5, and 7 days of contact with SA-G degradation products. Error bars represent the standard deviation. SA-G: sodium alginate–gelatin.

3.5. Printability Assessment

Once it was verified that the SA-G bioink met the rheological requirements needed for the extrusion-based process, filament formation and printing accuracy were assessed.

The first test consisted in printing four lines at different speeds and pressures with a 22G nozzle to evaluate filament continuity and the average line width using an optical microscope (Figure 7a). At all applied pressures, the average line width decreased as the speed increased (Figure 7b), achieving an average line width of 0.45 ± 0.124 mm at 25 kPa and of 0.5 ± 0.2 mm at 30 kPa for a printing speed of 20 mm/s, thus approximating, in both cases, the value of the nozzle inner diameter. The printing of narrower lines is due to a smaller amount of deposited material when higher speeds are applied. However, these speeds led to filament discontinuity and inhomogeneities. In contrast, filaments printed at 5 and 10 mm/s were continuous and homogeneous. Applying a higher pressure resulted in the deposition of a greater amount of material, as it induced higher flow rates. Therefore, at increasing pressures, broader lines were obtained. For instance, at a printing speed of 10 mm/s, the average line width was 0.592 ± 0.079 and 0.7 ± 0.1 for a pressure of 25 and 30 kPa, respectively. Thus, when using the 22G nozzle, the best compromise between printing resolution and filament continuity was achieved at a printing speed of 10 mm/s and an applied pressure of 25 kPa. Finally, the effect of nozzle diameter on printing resolution was evaluated by repeating the test using a 25G conical nozzle. In this case, to obtain a continuous filament, a minimum pressure of 120 kPa, nearly five times the pressure needed with the 22G nozzle, was required. Consistent with the results obtained using a 22G nozzle, the average line width decreased with increasing speed (Figure S1). The 25G nozzle provided higher printing resolution compared with the 22G nozzle, achieving an average line width of 0.30 ± 0.091 at a printing speed of 20 mm/s. These findings indicate that nozzle diameter significantly impacts the printing conditions and resolution. The choice of nozzle diameter should be tailored to the desired application. For bioinks containing cells, and especially primary cells, it is crucial to minimize cell death by optimizing the printing conditions. According to the literature, increased dispensing pressure is a major cause of cell damage due to elevated shear stresses on the cells [38]. As future studies will focus on assessing the formation of a functional photoreceptor layer that includes primary stem cells within the SA-G bioink, the 22G nozzle, a printing speed of 10 mm/s, and a pressure of 25 kPa were selected for further printability tests and for 3D bioprinting.

Printing accuracy was investigated through the filament fusion test, calculation of the printability index, and the comparison of real dimensions with the digital ones. Based on the results of the filament fusion test, filaments merged only when printed at a filament spacing of 0.5 mm (Figure 7c). The same outcome was obtained when printing Nivea Creme, considered in the literature as a demonstration ink, proving SA-G bioink's ability to print fine details like small pores with sharp angles [39]. To better investigate such capabilities, small pore geometries were obtained by printing porous circular discs (Figure 7d). Consequently, the printability index (Pr), which depends on the pore perimeter and area, was derived. The average Pr (1.080 ± 0.0025) was close to the ideal Pr, showing a high geometric accuracy. Ultimately, no relevant difference was encountered between the average diameter of the printed constructs (30.25 ± 0.056 mm) and that of the digital model (30 mm).

In conclusion, the SA-G bioink was demonstrated to be printable with high shape fidelity and printing accuracy.

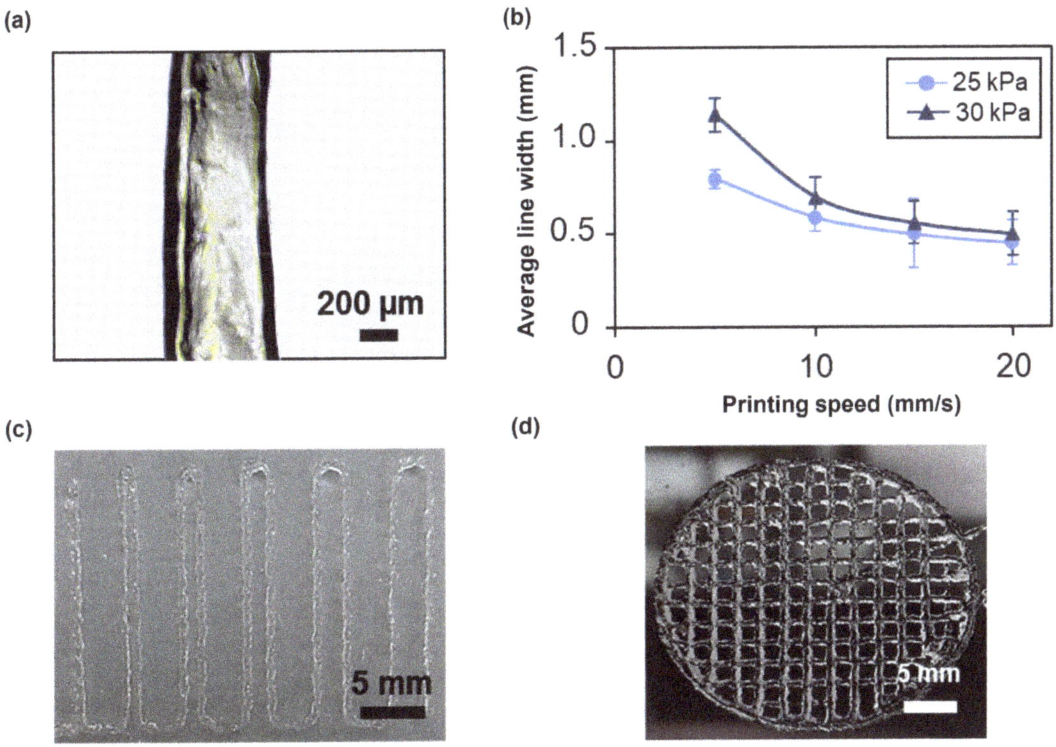

Figure 7. (**a**) Representative microscopy image of a single printed filament. Scale bar = 200 µm. (**b**) Average line width at different printing speeds and pressures using a 22G conical nozzle. Error bars represent the standard deviation. (**c**) Picture of the result of a filament fusion test. Scale bar = 5 mm. (**d**) Picture of porous circular discs. Scale bar = 5 mm. All the structures were printed using SA-G bioink. SA-G: sodium alginate–gelatin.

3.6. Bioprinting

Based on the results of the printability assessment (Section 3.5), the cell-laden SA-G bioink was printed at a speed of 10 mm/s and a pressure of 25 kPa. Post-printing cell viability was assessed by staining living (green) and dead (red) cells. High cell viability (>98%) was observed after 1 day of culture (Figure 8a,b) and was maintained at day 5 (Figure 8c), with no abnormal cell death detected. This is consistent with the in vitro cytotoxicity findings (Section 3.4), confirming that the SA-G bioink does not negatively affect cell viability. In addition, the high post-printing viability indicates that the shear stresses generated during bioink printing did not cause major cell death. Ultimately, Figure 8 demonstrates a homogeneous cell distribution throughout the printed constructs.

These results pave the way for incorporating more relevant cell types, such as primary stem cells, into the bioink. In this study, a well-established immortalized cell line (L929 cells) was selected for preliminary biocompatibility screening, as there is no photoreceptor cell line able to comprehensively recapitulate photoreceptor phenotypes [4]. Further investigations are needed to assess the impact of the SA-G bioink on the behavior of primary cells.

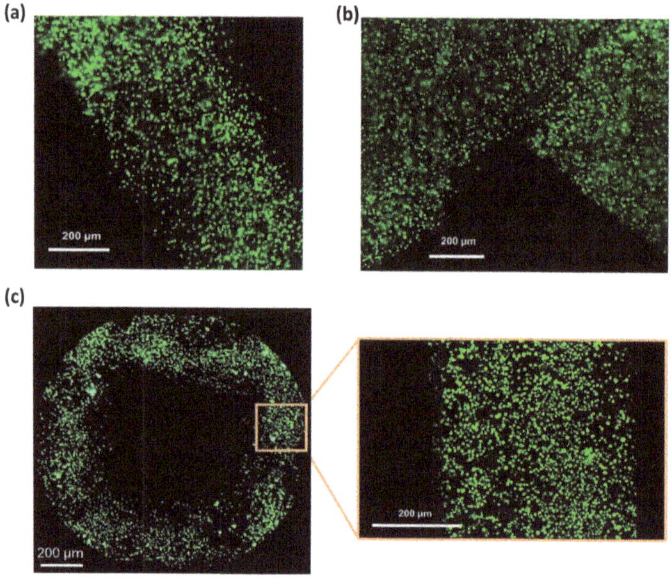

Figure 8. Fluorescence microscopy images of the LIVE/DEAD assay. Green and red indicate live and dead cells, respectively: (**a**) Image of a printed filament 1 day after printing; (**b**) Image of a corner within the printed grid 1 day after printing; (**c**) View of a pore of the printed grid and a zoomed-in view of a filament 5 days after printing. Scale bar = 200 μm.

4. Conclusions

In this study, we aimed to optimize an SA-G bioink formulation to match the properties of the native photoreceptor layer and to be successfully extruded and printed with good resolution. To assess its suitability as a substrate for photoreceptors in in vitro culture, we conducted a comprehensive characterization of an SA-G bioink. Our results showed that the stiffness of the bioink closely mimics the in vivo photoreceptor stiffness. This parameter plays a crucial role in the development of a functional photoreceptor layer, as it influences the cell behavior guiding the cell phenotype and genotype. In addition, through rheological and printability tests, we demonstrated the bioink's ability to fabricate multilayered structures characterized by high shape fidelity. This feature is crucial for obtaining a microstructure able to drive photoreceptor polarization, thus replicating their highly organized in vivo configuration. Ultimately, the viability of fibroblasts, both in contact with material extracts and embedded within the bioink, proved that the SA-G material is non-cytotoxic. Taken together, these findings suggest that the proposed SA-G bioink may be used as a tissue mimetic to produce a cellularized construct with the specific pore size and geometry capable of promoting photoreceptor development, alignment, and functional culture in vitro. Future studies should focus on incorporating stem cells into the SA-G bioink to form a functional photoreceptor layer in vitro. This step will require the selection of the appropriate culture medium and growth factors and the execution of several tests to determine the optimal cell density for mimicking the native high packing density of photoreceptors. Different printing patterns will then be explored to determine which microstructure best supports native-like spatial cell organization. The formation of a functional photoreceptor layer will be evaluated by conducting a thorough biological analysis, including of cell viability and proliferation and the expression of specific markers. Finally, in vivo studies need to be performed to assess the functional integration and stability of the engineered photoreceptor layer.

Supplementary Materials: The following supporting information can be downloaded at: https://www.mdpi.com/article/10.3390/app14166980/s1, Figure S1: Average line width at different printing speeds using a 25G conical nozzle.

Author Contributions: Conceptualization, B.B., S.M. and F.B.; methodology, B.B. and F.I.; investigation, B.B. and F.I.; data curation, B.B. and F.P.; writing—original draft preparation, B.B.; writing—review and editing, S.M., F.I., F.P., F.B. and M.L. All authors have read and agreed to the published version of the manuscript.

Funding: This research received no external funding.

Data Availability Statement: The raw data supporting the conclusions of this article will be made available by the authors on request.

Conflicts of Interest: The authors declare no conflict of interest.

References

1. Bovi dos Santos, G.; de Lima-Vasconcellos, T.H.; Móvio, M.I.; Birbrair, A.; Del Debbio, C.B.; Kihara, A.H. New Perspectives in Stem Cell Transplantation and Associated Therapies to Treat Retinal Diseases: From Gene Editing to 3D Bioprinting. *Stem Cell Rev. Rep.* **2024**, *20*, 722–737. [CrossRef] [PubMed]
2. Rajendran Nair, D.S.; Seiler, M.J.; Patel, K.H.; Thomas, V.; Camarillo, J.C.M.; Humayun, M.S.; Thomas, B.B. Tissue Engineering Strategies for Retina Regeneration. *Appl. Sci.* **2021**, *11*, 2154. [CrossRef] [PubMed]
3. Belgio, B.; Salvetti, A.P.; Mantero, S.; Boschetti, F. The Evolution of Fabrication Methods in Human Retina Regeneration. *Appl. Sci.* **2021**, *11*, 4102. [CrossRef]
4. Schnichels, S.; Paquet-Durand, F.; Löscher, M.; Tsai, T.; Hurst, J.; Joachim, S.C.; Klettner, A. Retina in a Dish: Cell Cultures, Retinal Explants and Animal Models for Common Diseases of the Retina. *Prog. Retin. Eye Res.* **2020**, *81*, 100880. [CrossRef]
5. Murphy, A.R.; Truong, Y.B.; O'Brien, C.M.; Glattauer, V. Bio-Inspired Human In Vitro Outer Retinal Models: Bruch's Membrane and Its Cellular Interactions. *Acta Biomater.* **2020**, *104*, 1–16. [CrossRef] [PubMed]
6. Steedman, M.R.; Tao, S.L.; Klassen, H.; Desai, T.A. Enhanced Differentiation of Retinal Progenitor Cells Using Microfabricated Topographical Cues. *Biomed. Microdevices* **2010**, *12*, 363–369. [CrossRef]
7. Jung, Y.H.; Phillips, M.J.; Lee, J.; Xie, R.; Ludwig, A.L.; Chen, G.; Zheng, Q.; Kim, T.J.; Zhang, H.; Barney, P.; et al. 3D Microstructured Scaffolds to Support Photoreceptor Polarization and Maturation. *Adv. Mater.* **2018**, *30*, e1803550. [CrossRef]
8. Hunt, N.C.; Hallam, D.; Karimi, A.; Mellough, C.B.; Chen, J.; Steel, D.H.W.; Lako, M. 3D Culture of Human Pluripotent Stem Cells in RGD-Alginate Hydrogel Improves Retinal Tissue Development. *Acta Biomater.* **2017**, *49*, 329–343. [CrossRef]
9. Ghareeb, A.E.; Lako, M.; Steel, D.H. Coculture Techniques for Modeling Retinal Development and Disease, and Enabling Regenerative Medicine. *Stem Cells Transl. Med.* **2020**, *9*, 1531–1548. [CrossRef]
10. McUsic, A.C.; Lamba, D.A.; Reh, T.A. Guiding the Morphogenesis of Dissociated Newborn Mouse Retinal Cells and HES Cell-Derived Retinal Cells by Soft Lithography-Patterned Microchannel PLGA Scaffolds. *Biomaterials* **2012**, *33*, 1396–1405. [CrossRef]
11. Lorber, B.; Hsiao, W.K.; Hutchings, I.M.; Martin, K.R. Adult Rat Retinal Ganglion Cells and Glia Can Be Printed by Piezoelectric Inkjet Printing. *Biofabrication* **2014**, *6*, 015001. [CrossRef] [PubMed]
12. Masaeli, E.; Forster, V.; Picaud, S.; Karamali, F.; Nasr-Esfahani, M.H.; Marquette, C. Tissue Engineering of Retina through High Resolution 3-Dimensional Inkjet Bioprinting. *Biofabrication* **2020**, *12*, 025006. [CrossRef] [PubMed]
13. Shi, P.; Tan, Y.S.E.; Yeong, W.Y.; Li, H.Y.; Laude, A. A Bilayer Photoreceptor-Retinal Tissue Model with Gradient Cell Density Design: A Study of Microvalve-Based Bioprinting. *J. Tissue Eng. Regen. Med.* **2018**, *12*, 1297–1306. [CrossRef] [PubMed]
14. Tripathi, S.; Mandal, S.S.; Bauri, S.; Maiti, P. 3D Bioprinting and Its Innovative Approach for Biomedical Applications. *MedComm* **2023**, *4*, e194. [CrossRef] [PubMed]
15. Perin, F.; Spessot, E.; Famà, A.; Bucciarelli, A.; Callone, E.; Mota, C.; Motta, A.; Maniglio, D. Modeling a Dynamic Printability Window on Polysaccharide Blend Inks for Extrusion Bioprinting. *ACS Biomater. Sci. Eng.* **2023**, *9*, 1320–1331. [CrossRef] [PubMed]
16. Schwab, A.; Levato, R.; D'Este, M.; Piluso, S.; Eglin, D.; Malda, J. Printability and Shape Fidelity of Bioinks in 3D Bioprinting. *Chem. Rev.* **2020**, *120*, 11028–11055. [CrossRef] [PubMed]
17. Shi, P.; Edgar, T.Y.S.; Yeong, W.Y.; Laude, A. Hybrid Three-Dimensional (3D) Bioprinting of Retina Equivalent for Ocular Research. *Int. J. Bioprint.* **2017**, *3*, 138–146. [CrossRef] [PubMed]

18. Wang, P.; Li, X.; Zhu, W.; Zhong, Z.; Moran, A.; Wang, W.; Zhang, K.; Chen, S. 3D Bioprinting of Hydrogels for Retina Cell Culturing. *Bioprinting* **2018**, *12*, e00029. [CrossRef] [PubMed]
19. Hiller, T.; Berg, J.; Elomaa, L.; Röhrs, V.; Ullah, I.; Schaar, K.; Dietrich, A.C.; Al-Zeer, M.A.; Kurtz, A.; Hocke, A.C.; et al. Generation of a 3D Liver Model Comprising Human Extracellular Matrix in an Alginate/Gelatin-Based Bioink by Extrusion Bioprinting for Infection and Transduction Studies. *Int. J. Mol. Sci.* **2018**, *19*, 3129. [CrossRef]
20. Fantini, V.; Bordoni, M.; Scocozza, F.; Conti, M.; Scarian, E.; Carelli, S.; Di Giulio, A.M.; Marconi, S.; Pansarasa, O.; Auricchio, F.; et al. Bioink Composition and Printing Parameters for 3D Modeling Neural Tissue. *Cells* **2019**, *8*, 830. [CrossRef]
21. Bedell, M.L.; Navara, A.M.; Du, Y.; Zhang, S.; Mikos, A.G. Polymeric Systems for Bioprinting. *Chem. Rev.* **2020**, *120*, 10744–10792. [CrossRef] [PubMed]
22. Shams, E.; Barzad, M.S.; Mohamadnia, S.; Tavakoli, O.; Mehrdadfar, A. A Review on Alginate-Based Bioinks, Combination with Other Natural Biomaterials and Characteristics. *J. Biomater. Appl.* **2022**, *37*, 355–372. [CrossRef] [PubMed]
23. Wang, X.; Ao, Q.; Tian, X.; Fan, J.; Tong, H.; Hou, W.; Bai, S. Gelatin-Based Hydrogels for Organ 3D Bioprinting. *Polymers* **2017**, *9*, 401. [CrossRef] [PubMed]
24. Fayyazbakhsh, F.; Khayat, M.J.; Leu, M.C. 3D-Printed Gelatin-Alginate Hydrogel Dressings for Burn Wound Healing: A Comprehensive Study. *Int. J. Bioprint.* **2022**, *8*, 274–291. [CrossRef] [PubMed]
25. Lu, H.; Butler, J.A.; Britten, N.S.; Venkatraman, P.D.; Rahatekar, S.S. Natural Antimicrobial Nano Composite Fibres Manufactured from a Combination of Alginate and Oregano Essential Oil. *Nanomaterials* **2021**, *11*, 2062. [CrossRef] [PubMed]
26. Lee, B.H.; Lum, N.; Seow, L.Y.; Lim, P.Q.; Tan, L.P. Synthesis and Characterization of Types A and B Gelatin Methacryloyl for Bioink Applications. *Materials* **2016**, *9*, 797. [CrossRef] [PubMed]
27. Belgio, B.; Boschetti, F.; Mantero, S. Towards an In Vitro Retinal Model to Study and Develop New Therapies for Age-Related Macular Degeneration. *Bioengineering* **2021**, *8*, 18. [CrossRef] [PubMed]
28. Rahimnejad, M.; Labonté-Dupuis, T.; Demarquette, N.R.; Lerouge, S. A Rheological Approach to Assess the Printability of Thermosensitive Chitosan-Based Biomaterial Inks. *Biomed. Mater.* **2020**, *16*, 015003. [CrossRef]
29. Ribeiro, A.; Blokzijl, M.M.; Levato, R.; Visser, C.W.; Castilho, M.; Hennink, W.E.; Vermonden, T.; Malda, J. Assessing Bioink Shape Fidelity to Aid Material Development in 3D Bioprinting. *Biofabrication* **2018**, *10*, 014102. [CrossRef]
30. Bociaga, D.; Bartniak, M.; Grabarczyk, J.; Przybyszewska, K. Sodium Alginate/Gelatine Hydrogels for Direct Bioprinting-the Effect of Composition Selection and Applied Solvents on the Bioink Properties. *Materials* **2019**, *12*, 2669. [CrossRef]
31. Li, Z.; Huang, S.; Liu, Y.; Yao, B.; Hu, T.; Shi, H.; Xie, J.; Fu, X. Tuning Alginate-Gelatin Bioink Properties by Varying Solvent and Their Impact on Stem Cell Behavior. *Sci. Rep.* **2018**, *8*, 8020. [CrossRef]
32. Barceló, X.; Eichholz, K.F.; Garcia, O.; Kelly, D.J. Tuning the Degradation Rate of Alginate-Based Bioinks for Bioprinting Functional Cartilage Tissue. *Biomedicines* **2022**, *10*, 1621. [CrossRef]
33. Ferrara, M.; Lugano, G.; Sandinha, M.T.; Kearns, V.R.; Geraghty, B.; Steel, D.H.W. Biomechanical Properties of Retina and Choroid: A Comprehensive Review of Techniques and Translational Relevance. *Eye* **2021**, *35*, 1818–1832. [CrossRef] [PubMed]
34. He, J.; Chen, R.; Lu, Y.; Zhan, L.; Liu, Y.; Li, D.; Jin, Z. Fabrication of Circular Microfluidic Network in Enzymatically-Crosslinked Gelatin Hydrogel. *Mater. Sci. Eng. C* **2016**, *59*, 53–60. [CrossRef] [PubMed]
35. Qu, Y.; He, Y.; Zhang, Y.; Ma, T.; Zhu, J.; Miao, Y.; Dai, C.; Humayun, M.; Zhou, Q.; Chen, Z. Quantified Elasticity Mapping of Retinal Layers Using Synchronized Acoustic Radiation Force Optical Coherence Elastography. *Biomed Opt. Express.* **2018**, *9*, 4054–4063. [CrossRef]
36. Jain, T.; Baker, H.B.; Gipsov, A.; Fisher, J.P.; Joy, A.; Kaplan, D.S.; Isayeva, I. Impact of Cell Density on the Bioprinting of Gelatin Methacrylate (GelMA) Bioinks. *Bioprinting* **2021**, *22*, e00131. [CrossRef]
37. Ouyang, L.; Highley, C.B.; Rodell, C.B.; Sun, W.; Burdick, J.A. 3D Printing of Shear-Thinning Hyaluronic Acid Hydrogels with Secondary Cross-Linking. *ACS Biomater. Sci. Eng.* **2016**, *2*, 1743–1751. [CrossRef] [PubMed]
38. Xu, H.Q.; Liu, J.C.; Zhang, Z.Y.; Xu, C.X. A Review on Cell Damage, Viability, and Functionality during 3D Bioprinting. *Mil. Med. Res.* **2022**, *9*, 70. [CrossRef]
39. Paxton, N.; Smolan, W.; Böck, T.; Melchels, F.; Groll, J.; Jungst, T. Proposal to Assess Printability of Bioinks for Extrusion-Based Bioprinting and Evaluation of Rheological Properties Governing Bioprintability. *Biofabrication* **2017**, *9*, 044107. [CrossRef]

applied sciences

MDPI

Article

Stress Analysis on the Ankle Joint during Incline and Decline Standing

Noor Arifah Azwani Abdul Yamin [1], Khairul Salleh Basaruddin [1,2,*], Muhammad Farzik Ijaz [3,*], Mohd Hanafi Mat Som [2,4], Muhammad Nazrin Shah Shahrol Aman [5] and Hiroshi Takemura [6]

1 Faculty of Mechanical Engineering & Technology, Universiti Malaysia Perlis, Arau 02600, Malaysia; ariifahazwani@studentmail.unimap.edu.my
2 Medical Devices and Health Sciences, Sports Engineering Research Center (SERC), Universiti Malaysia Perlis, Arau 02600, Malaysia; mhanafi@unimap.edu.my
3 Mechanical Engineering Department, College of Engineering, King Saud University, Riyadh 11421, Saudi Arabia
4 Faculty of Electronic Engineering & Technology, Universiti Malaysia Perlis, Arau 02600, Malaysia
5 Faculty of Electrical Engineering & Technology, Universiti Malaysia Perlis, Arau 02600, Malaysia; nazrinshahrol@unimap.edu.my
6 Department of Mechanical Engineering, Faculty of Science and Technology, Tokyo University of Science, Chiba 278-8510, Japan; takemura@rs.tus.ac.jp
* Correspondence: khsalleh@unimap.edu.my (K.S.B.); mijaz@ksu.edu.sa (M.F.I.)

Abstract: In daily routine movement, the ankle joint plays a crucial role in stability and mobility, especially when different types of terrain are involved. However, the simple task of standing can become a biomechanical difficulty when performed on a slope since demands that have to be accommodated are made on the complex structure of the ankle joint. The purpose of this study is to develop finite element (FE) models of the ankle joint with different inclined foot postures and to analyse the stress distributions on the ankle joint while standing on an inclined or declined surface. In this study, the FE model of the foot was developed, and von Mises stress distribution at the ankle joint was explored. The results show that the bone, cartilage, and ligament of the ankle experienced a different von Mises stress distribution pattern during flat standing in comparison with slope standing. In addition, this study found that the maximum von Mises stress distribution at the component of the ankle joint is higher during slope standing than flat standing. Taken together, these results suggest that slope standing, both inclined and declined, with more than 10° inclination, might contribute to a higher risk of injury as a higher maximum stress was observed. Therefore, to maintain proper body posture, it is suggested that weight be evenly distributed at both feet, since this can reduce stress at the ankle.

Keywords: surface inclination; stability; finite element analysis; incline standing

1. Introduction

Standing on slopes, whether inclined or declined, presents unique challenges to the human body's balance and stability [1]. In everyday life, encounters with sloped surfaces are common, ranging from walking up a hill to navigating stairs or ramps. The human body's intricate interactions with muscle control, gravitational forces, and sensory systems enable it to stay balanced on sloped surfaces. When standing on an inclined or declined surface, individuals must adapt their posture, muscle activation patterns, and weight distribution to counteract the effects of gravity pulling them downhill or uphill [2]. This adjustment involves complex neuromuscular coordination to stabilise the body and prevent falls or loss of balance.

One of the benefits of standing on an inclined or declined surface in biomechanics is that it challenges the body's proprioceptive system [3], which is responsible for sensing and controlling body position and movement. This can lead to improved coordination and proprioception, ultimately enhancing overall stability and balance. Additionally, standing on an inclined or declined surface can improve muscle strength and endurance [4], as well as increase blood flow to the lower extremities. This can also help to improve posture and reduce muscle soreness.

However, standing on an inclined or declined surface also poses certain risks in biomechanics. The uneven surface can place added stress on the joints, particularly the ankles and knees, increasing the risk of injury [5]. It is important to start with a gradual incline or decline and progress slowly to allow the body to adapt and strengthen the necessary muscles and ligaments. It is also important to use proper form when standing on an incline or decline. Bending the knees and keeping the body upright is essential. Finally, it is important to stretch before and after any exercise on an incline or decline to reduce the risk of injury.

Ankle stability and strength play a crucial role when standing on an inclined or declined surface in biomechanics. The uneven terrain can put extra strain on the ankles, increasing the risk of sprains or other injuries [6]. It is important to have good ankle mobility and engage the muscles surrounding the ankle joint to maintain balance and prevent instability. Ankle exercises such as squats, lunges, and calf raises can help to improve stability and strength.

The finite element analysis (FEA) has been utilised as an addition to experimental methods in order to anticipate the load distribution between the foot and various supports. FEA was commonly utilised previously in order to investigate the ankle joint. A model of the foot using FEA has been developed to investigate the dynamic behaviour and internal loading during standing [7]. In addition, the ankle joint model using three-dimensional (3D) actual geometry of skeletal and soft tissue has also been developed in order to analyse the effect of soft tissue stiffness on plantar pressure distribution and internal load transfer [8]. Instead of static, FEA has also been integrated with rigid body dynamic analysis during gait to obtain the contact pressure distribution at the knee and ankle joint [9]. Other than that, in the physical well-being aspect, the FEA model has been established in order to explore the effect of sagittal angle related to posterior malleolus fracture [10]. FEA has also been applied in developing a new approach to ankle prosthesis [11] and in investigating the performance of the ankle prosthesis during gait [12]. These approaches provide additional information on the internal stresses and strains of the ankle–foot complex [8]. A simulation on finite element analysis is required to investigate and analyse the ankle joint due to these conditions. However, there is a limited research focus on the ankle joint model simulation during slope standing.

It is crucial to comprehend the physiological reactions and biomechanical theories linked to slope standing in order to improve sports performance, optimise rehabilitation regimens, and build ergonomic workspaces. Researchers and practitioners may develop interventions to lessen the difficulties associated with incline and decline standing by clarifying the elements impacting balance and stability on sloped surfaces. This will ultimately improve safety, effectiveness, and well-being in a variety of contexts. Thus, the objectives of this study are to develop finite element models of the ankle joint with different inclined foot postures and to analyse the stress distributions on the ankle joint while standing on an inclined or declined surface.

2. Materials and Methods

The workflow of the present method is shown in Figure 1. A three-dimensional (3D) model of the ankle was developed based on the reference model obtained from the Turbosquid database. The model represents a foot of a male subject with an age of 23 years, body weight of 60 kg, and foot length of 245 cm. The 3D model was created using Solidworks 2018 and meshed using ANSYS version 19.2. A convergence test was performed to determine the optimum element size. The meshed model was then put through the common FE analysis process.

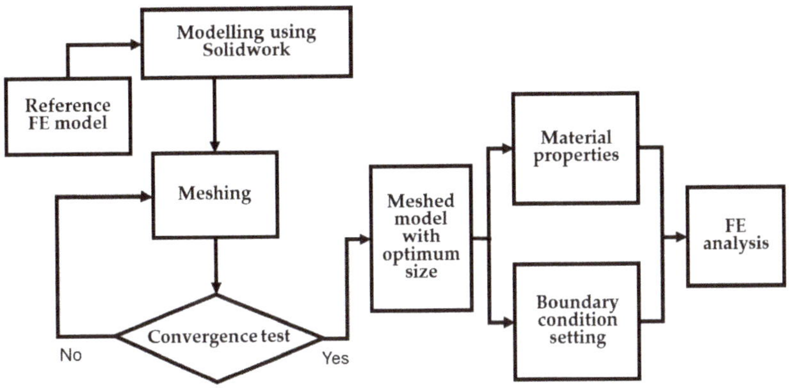

Figure 1. Workflow of the present study.

2.1. Geometrical Modelling of the Ankle Joint

Figure 2 shows a geometrical model of bone structure in 3D. The model is not combined as one model because there are no ligaments and cartilages in the original foot model. Hence, the gaps between the bones need to be filled with the cartilage and connected by ligaments. The phalanges were assumed to contact the plane surface.

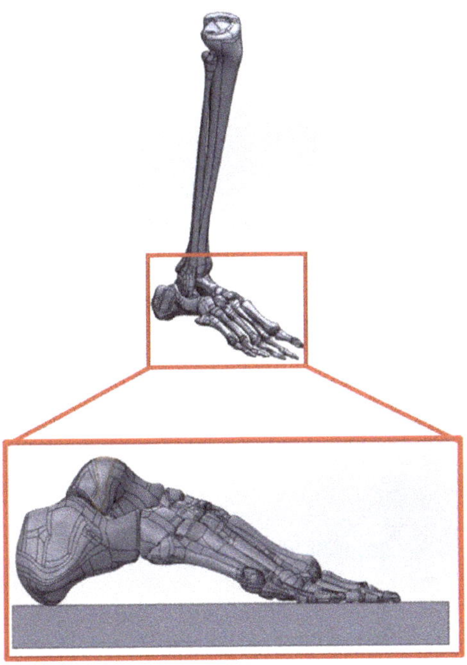

Figure 2. The bone structure of the foot on a plane surface.

The construction and insertion site of the cartilage and ligaments to the foot model are based on the X-ray image of the subject's foot. Cartilages were added inside the model using Solidworks 2018 software. Cartilage is a tissue that connects bone to bone. A simple shape was created and added at the specified location based on the X-ray image and verified through literature. By changing the plane's degree and axis, all cartilage can be created and connected in the model. The intersect function was used to create an inner volume between the bones. The interaction volume is the simulated cartilage in the model. Figure 3a shows the cartilage in the model, which is in red.

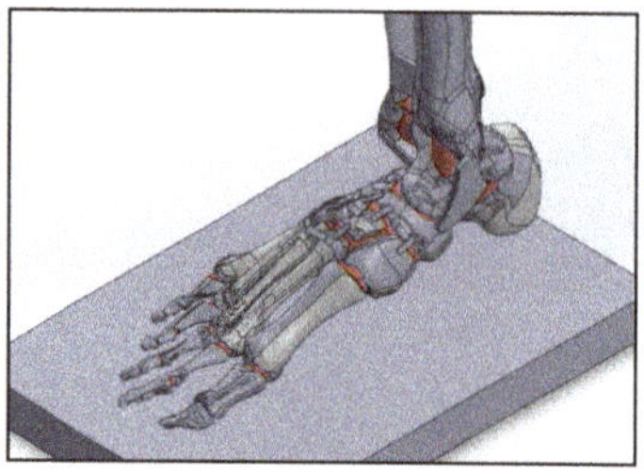

(a)

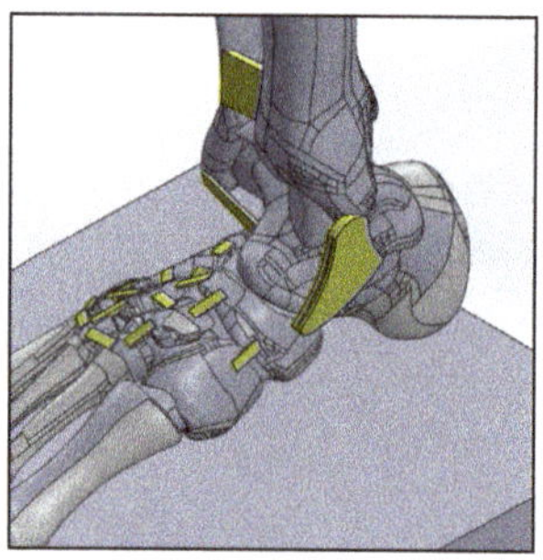

(b)

Figure 3. Model of foot. (**a**) Cartilages in ankle joint model. (**b**) Ligaments in the foot.

Next, the ligaments were included in the model. A simple structure was created and attached to all the important locations to simulate the ligaments. A model that included ligaments (yellow) was created, as shown in Figure 3b. After the model was created, this model was simulated to stand on the level surface. For various inclined and declined surfaces, the plate angle was changed to simulate the slope surface standing. Figure 4 shows the different models in the various inclined surfaces created according to the literature [13]. The knee was assumed to adopt the effect of incline/decline surfaces to maintain the orientation of tibia perpendicular to the foot.

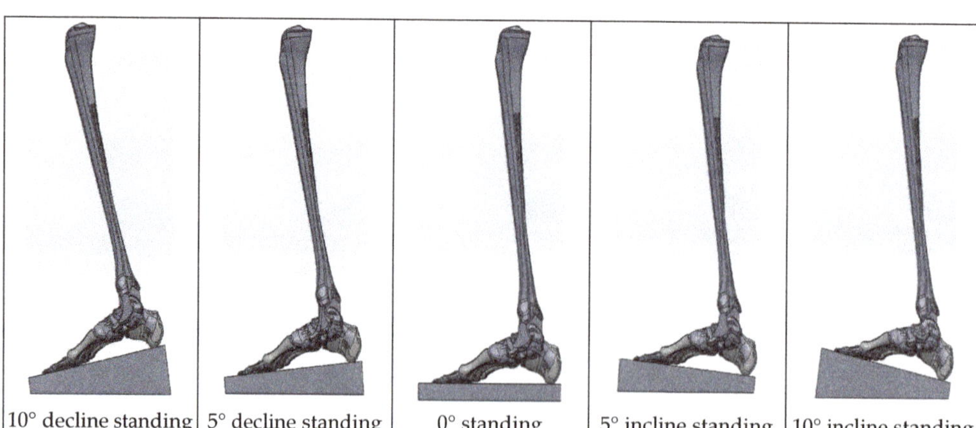

Figure 4. Models of the foot at various incline/decline surfaces.

2.2. Finite Element Model of Ankle Joint

(a) Meshing

Meshing is a part of the engineering simulation process where complex geometries are divided into elements. Good meshing is required in critical areas such as areas of interest in stress distribution (ankle joint) to increase the accuracy of simulation results. Before meshing was performed, the model was exported in a STEP file and opened in ANSYS. The finite element model was meshed with the ten-nodes 3D-tetrahedral element for bone, cartilage, ligament, and ground support. The ligaments were modelled as a solid element to observe the stress distribution on it. Although most of the previous studies used a tension-only truss to model the ligaments [8,11], there are also previous studies that used isotropic solid element, as mentioned by [14]. Two elements were set on the thickness of each ligament. The model was performed under linear static analysis. Figure 5 shows the complete meshing of the ankle joint model, and the magnified view of the ankle joint.

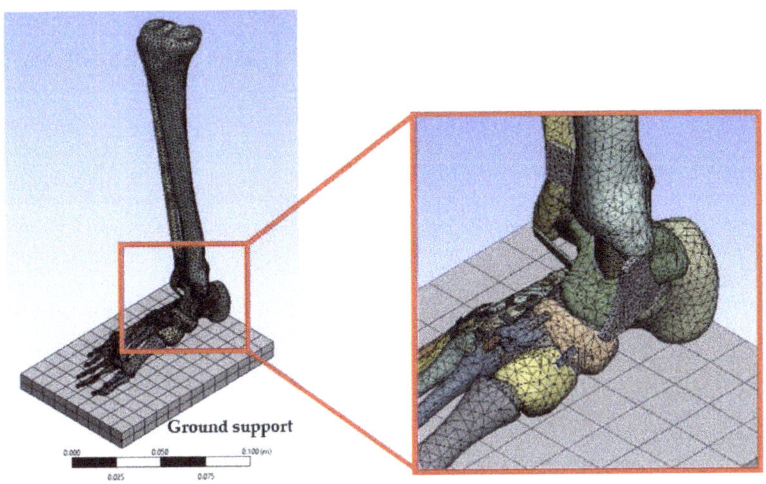

Figure 5. Complete meshed model with magnified view of the ankle joint.

(b) Convergence test

The number of elements was based on different sets of mesh sizes and types from the previous meshing. The meshing process of the model needs to be set with fixed mesh types but different sizes to obtain a different number of elements in each part. The solution made with a mesh discretization was observed and recorded. Then, it proceeded with finer mesh, and the results were compared with the previous test. If the results showed a different significant amount, the process was repeated with finer mesh. This step was repeated until the results were approximately similar for the mesh to be considered enough for the model. The increment of the element number resulted in a longer computational time due to the time-consuming nature of solving the mathematical models to complete the solution. Excessively large values of element size should be avoided because the result produced by ANSYS will be inaccurate and contain a higher percentage value of error.

A simulation with an arbitrary load of 300 N was applied on top of the tibia for the convergence test. The influence of mesh size on the maximum von Mises stress of the model was recorded and the meshing process was repeated with a decrement of the mesh size by 0.1 mm (from 2.5 mm to 1.5 mm) as shown in Figure 6. The changes in mesh size only involve the structure of bone while the cartilages and ligaments stay constant. Based on the convergence test curve shown, the optimum element size obtained is 2.0 mm with 200,485 elements, where the graph starts to reach a constant at that mesh size.

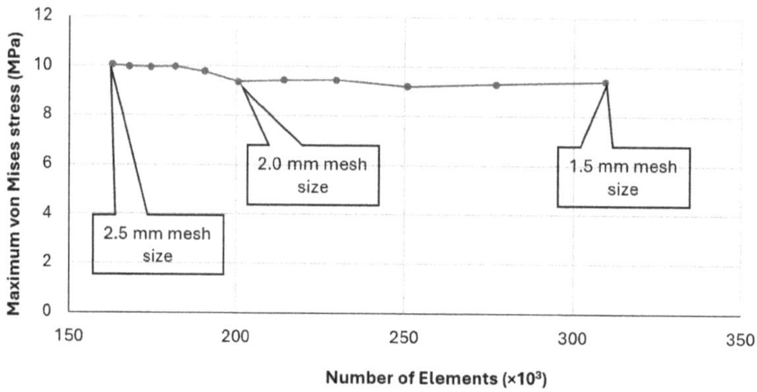

Figure 6. Convergence test results.

(c) Material properties

After the optimum mesh size was found, the material properties in each part of the ankle joint were defined. There were four central parts of material: bone, ligament, cartilage, and ground support. Each of these parts has different material properties. The material properties for the model can be directly imported from the engineering data of ANSYS. Bone, ligament, and cartilage have their own unique Young's Modulus and Poisson's ratio. However, there are no existing biological material properties such as ligament and cartilage in the engineering data. New material properties have to be added manually to the engineering data before defining these models. The material properties of bone, ligament, and cartilage are adopted from the literature [11] and tabulated in Table 1. The material properties were assigned to every component of the model to obtain accurate results.

Table 1. Material properties of bone, ligament, and cartilage [11].

Part	Young's Modulus (MPa)	Poisson's Ratio
Bone	7300	0.3
Ligament	260	0.4
Cartilage	10	0.4
Ground Support	200,000	0.29

2.3. Boundary Conditions

A simulation on the foot model with ankle joint was performed under static conditions. A vertical downward load of 300 N was applied on top of the tibia to reflect the self-weight of the subject during standing. Direction of the force was set to vertical to correspond to the gravitational force. The force was obtained approximately based on 60 kg of the subject's body weight and divided by two, to reflect double leg standing condition. The rigid ground support was set as fixed. The foot model was attached to the ground support top surface. The type of contact was set as bonded for all connections of bone, ligament, cartilage, and ground support. These boundary conditions will be similar for all models with various inclined slopes. In this study, the subject is assumed in balance for all loading cases. Hence, the motion of the ankle joint in eversion/inversion that may occur to keep the subject in balance was negligible.

2.4. Prediction of Failure

The result obtained was used to determine the location of possible failure that might occur at the ankle joint. The location for peak stress that leads to injury during incline or decline standing can be identified by checking the peak stress and strain of the model. Failure possibility has been checked using the equation of safety, Factor of Safety (FOS).

$$\text{FOS} = \frac{\text{Allowable stress}}{\text{Maximum Von Mises Stress}} \quad (1)$$

The lower the percentage of FOS, the higher the chance of failure at the particular part (i.e., bone, ligaments, and cartilage) observed.

3. Results

3.1. Von Mises Stress Distribution at the Ankle Joint during Slope Standing

The FE model developed enables von Mises stress distribution to be predicted during level and slope standing. For validation of the model, the acquired von Mises stress distribution was compared with results given in the literature that were obtained during level standing on the talus, tibia, and fibula, as shown in Table 2. The Young Modulus and Poisson Ratio for bone, cartilage, and ligaments used in the current study is similar to the compared study in the literatures [11,15]. As seen in the table, the value of von Mises stress distributions is comparable with a slight difference. The slight difference might be due to different boundary conditions applied in developing the FE model.

Table 2. Comparison of maximum von Mises stress on ankle joint during level standing between present study and the literatures [11,15].

	Maximum von Mises Stress from the Present Study	Maximum von Mises Stress Distributions from Literature	
		Ozen et al. [11]	Mondal et al. [15]
Talus (MPa)	1.86	2.07	2.38
Tibia (MPa)	4.84	5.50	5.86
Fibula (MPa)	1.18	1.28	1.30

The FE model of the ankle on five inclination conditions was analysed and the von Mises stress distributions and maximum von Mises stress were obtained. The result of the ankle joint model was specified into the effect on bones, i.e., talus, tibia and fibula, cartilage, and ligament. The von Mises stress was analysed according to colour contour as represented in Tables 3 and 4.

Table 3. Von Misses stress distribution on bones (talus, tibia, and fibula).

Inclination	Von Mises Stress			
	Talus		Tibia	Fibula
	Superior View	Inferior View		
−10°	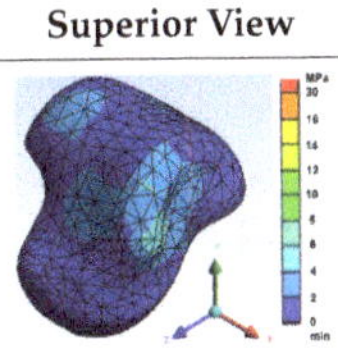	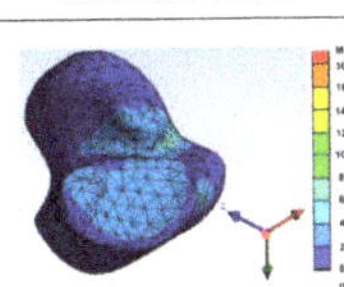	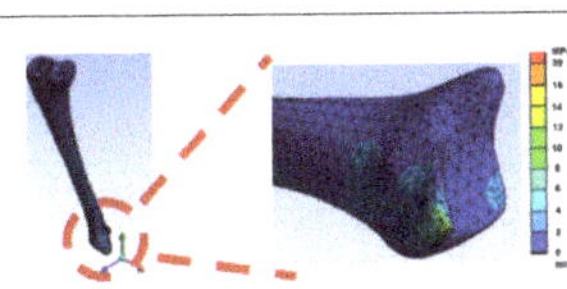	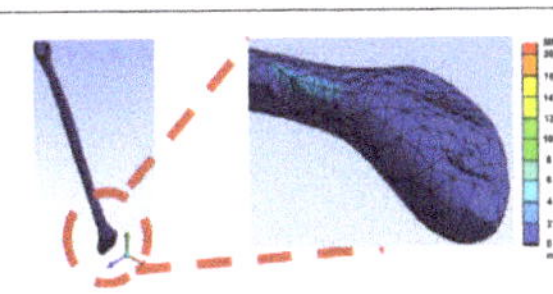
−5°	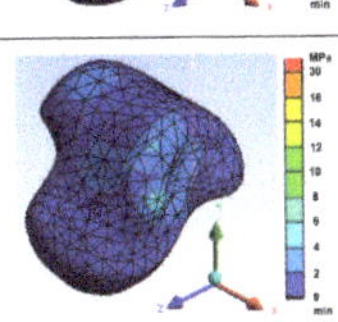	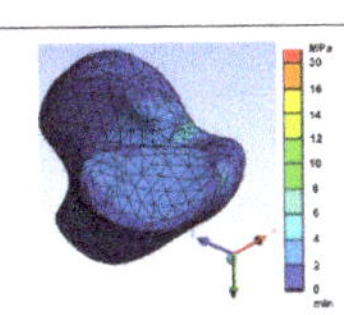	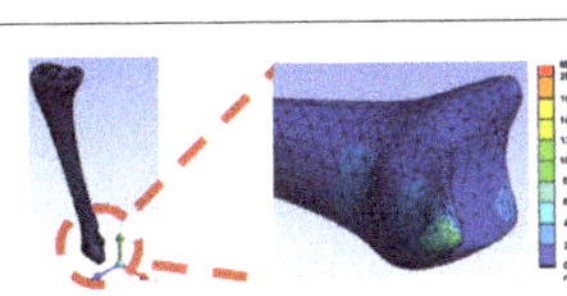	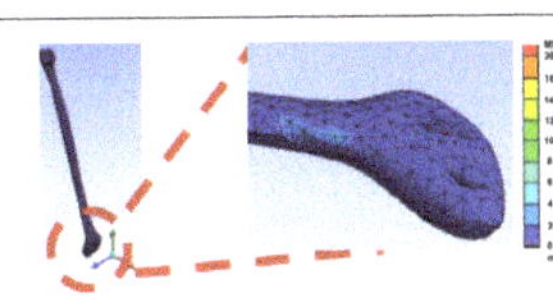

Table 3. *Cont.*

Inclination	Von Mises Stress			
	Talus		Tibia	Fibula
	Superior View	Inferior View		
0°	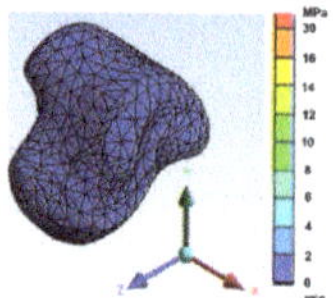	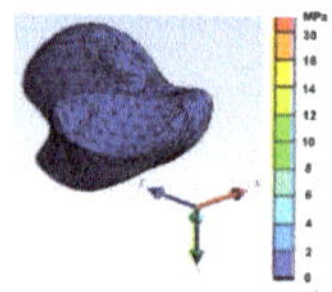	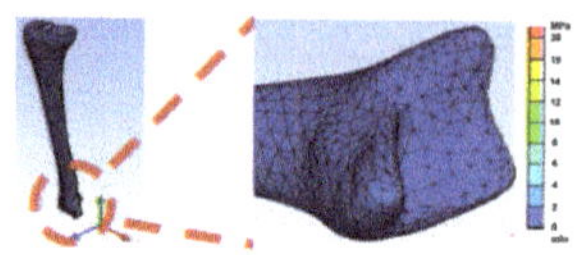	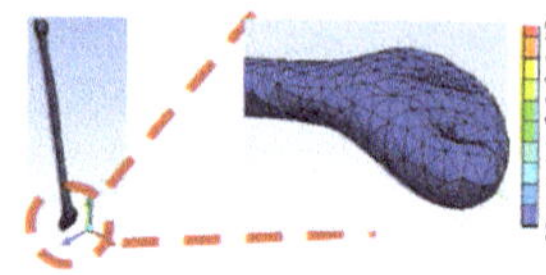
5°	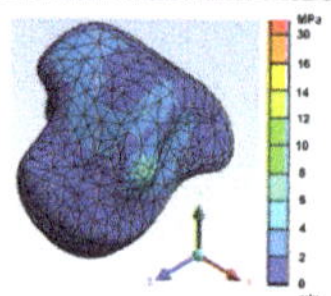	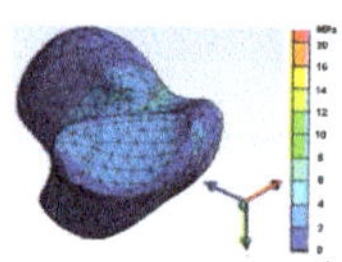	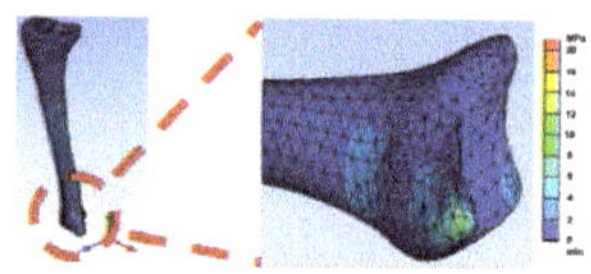	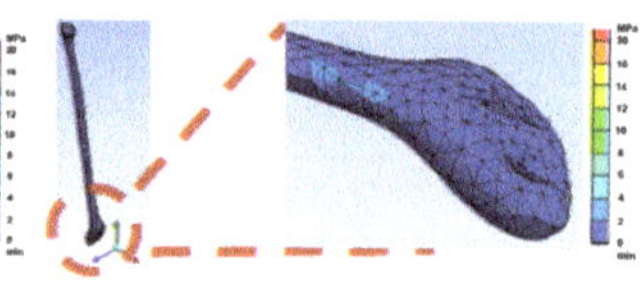
10°	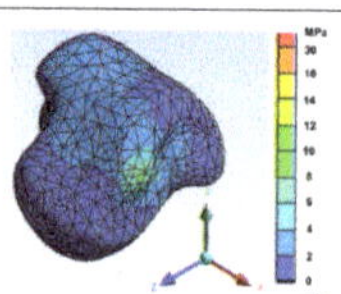	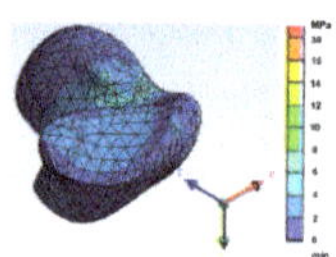	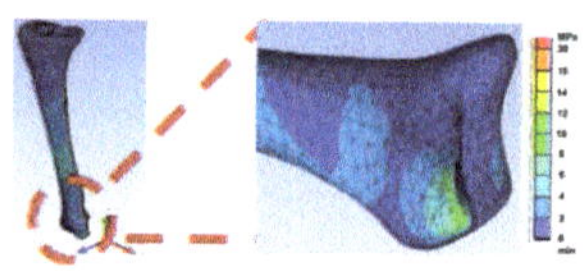	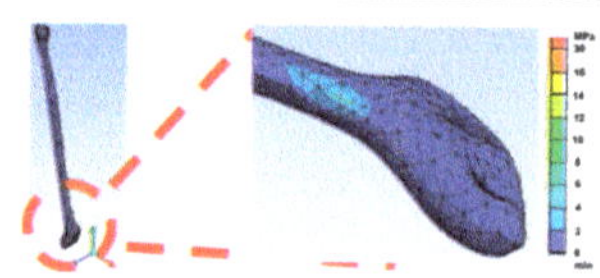

Table 4. Von Misses stress distribution on cartilage and ligament.

Surface Inclination Angle	Von Mises Stress	
	Cartilage	Ligament
−10°	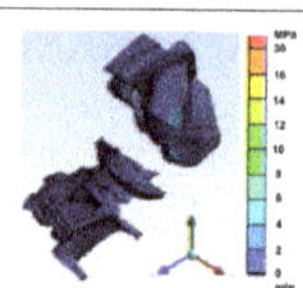	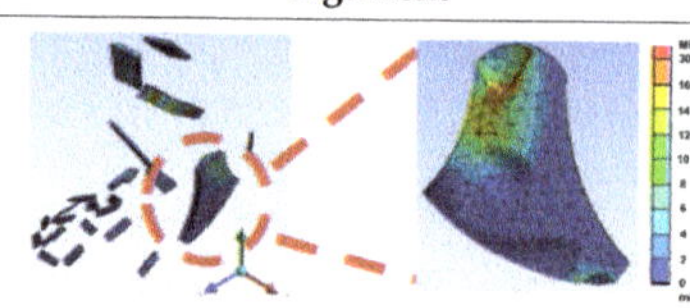
−5°	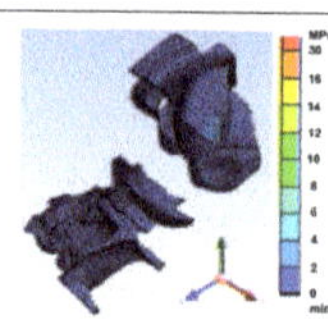	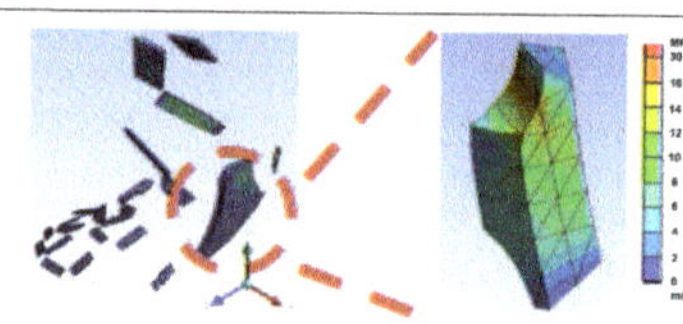
0°	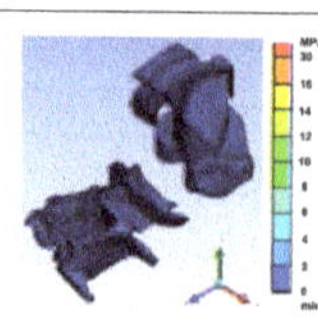	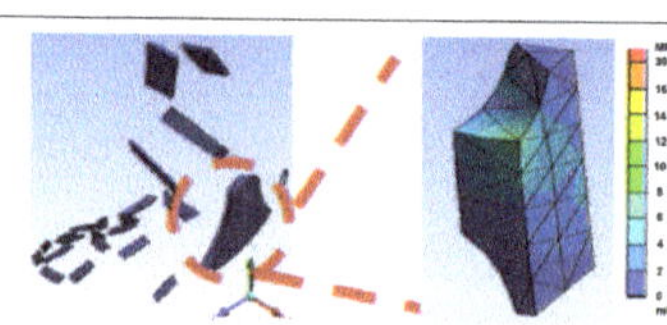
5°	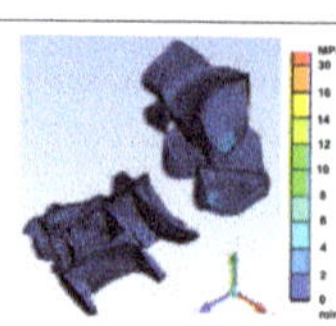	

Table 4. *Cont.*

Surface Inclination Angle	Von Mises Stress	
	Cartilage	Ligament
10°	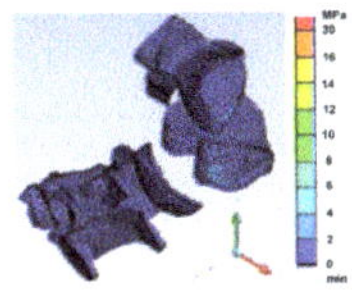	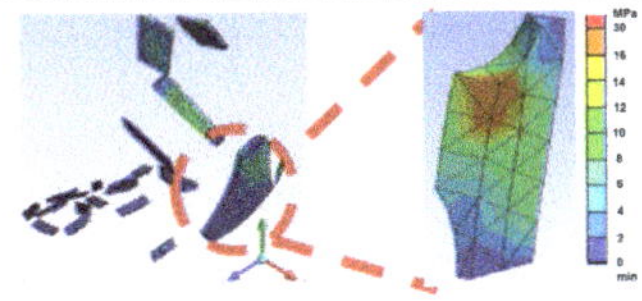

As can be seen in Table 3, an overall blue colour contour, which represents 0–2 MPa stress, has covered all of the talus during flat standing. Instead of that, a light blue contour, which represents 2–4 MPa, was spread around the talus, and a small lighter blue contour, representing 4–6 MPa, can be observed at the edge and bottom of the talus during 5° and 10° decline standing. As for 5° incline standing, similar to decline standing, the talus was covered with a light blue contour (2–4 MPa) but with a larger contour at the top of the talus, and there was a small lighter blue contour, which represents 4–6 MPa, that was located at the edge and bottom of the talus. During 10° incline standing, a light blue contour (2–4 MPa) was more significant than another standing covered on the top surface of the talus. The stress increased until the greenish-blue contour, which represents 6–8 MPa, appeared at the edge and bottom of the talus bone.

Other than that, the tibia was covered with an overall blue contour, representing 0–2 MPa, with a small light blue that represents 2–4 MPa at the bottom edge of the tibia during flat standing. Meanwhile, the tibia was covered with a light blue contour (2–4 MPa) around the middle part during the 5° decline standing and covered with a lighter blue contour (4–6 MPa) that covered the middle and bottom of the tibia during 10° decline standing. The contour was more significant than in other situations. A small green and yellow contour that represents 10–14 MPa was shown at the end edge of the tibia. Instead, there was an increment in von Mises stress at the middle and bottom of the tibia where a light blue contour was observed, which represents 2–4 MPa, during 5° incline standing. The stress increased at the bottom edge of the tibia as the colour changed from light blue to light green, rising from 4 MPa to 12 MPa. Similarly, the talus was covered with a light blue contour during 10° incline standing, but there was a yellow contour at the end edge of the tibia that represents 12–14 MPa.

In addition, as shown in Table 3, the whole fibula was covered with a blue contour (0–2 MPa) during flat standing. Instead, a small light blue contour (2–4 MPa) was observed at the bottom of the fibula during both 5° and 10° decline standing. A similar spot of light blue contour (2–4 MPa) was observed during 5° and 10° incline standing, but there was a tiny spot of lighter blue representing 4–6 MPa found during 10° incline standing.

Furthermore, Table 4 presents the von Mises stress distribution at the cartilage and ligament of the foot. From the table, it can be seen that the whole cartilage was covered by a blue contour, which represents 0–2 MPa von Mises stress during flat standing. However, for both 5° and 10° decline standing, a small light blue contour (2–4 MPa) was concentrated around the connected area of the tibia and talus as well as calcaneus and talus. A similar result was shown for 5° and 10° incline standing.

Similar to cartilage, the von Mises stress distribution on the ligament during flat standing was around 0–2 MPa as a blue contour covering the whole ligament was observed. However, during the 5° decline standing, a bright green contour, which represents 10–12 MPa, was observed at the talocalcaneal ligament which connects the talus and calcaneus. In addition, a light blue to green colour contour (8–10 MPa) was observed at the talofibular and tibiofibular ligament and an orange contour (16–30 MPa) occurred at the talocalcaneal ligament. Likewise, during 10° decline standing, small orange contours (16–30 MPa) were observed, but at the calcaneofibular ligament.

The von Mises distribution during incline standing at 5° inclination had a similar result to decline standing where a bright green contour (10–12 MPa) at the talocalcaneal ligament was shown. Meanwhile, during 10° incline standing, the green contour (8–10 MPa) was observed at the talofibular, tibiofibular, and deltoid ligaments.

3.2. Maximum von Misses Stress

The maximum von Mises stress on the talus, tibia, and fibula was also analysed in this study. The maximum von Mises stress found during level and slope standing is presented in Figure 7.

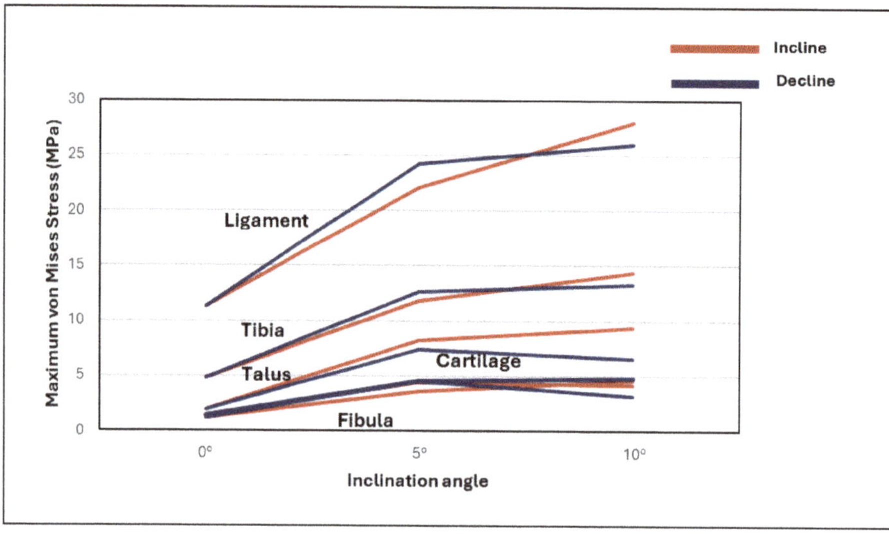

Figure 7. Comparison of maximum Von Mises stress during flat and slope standing.

Based on the result, it was reported that the ligament experienced an obvious increment of von Mises stress from 24.2 MPa on 5° to 26.0 MPa on 10° inclination compared to 11.3 MPa on flat standing during decline standing. In addition, the highest von Mises stress was found during standing on 10° inclination—28 MPa compared to 22.1 MPa on 5° inclination and 11.3 MPa on flat standing. Overall, the results show that the maximum von Mises stress at the ankle joint is higher during incline and decline standing compared to flat standing for all components of the ankle.

3.3. Failure Prediction

Based on Equation (1), the failure prediction that is determined by factor of safety (FOS) is presented in Figure 8. In general, healthy bones have an FOS greater than one, meaning they can withstand loads greater than those typically encountered without fracturing or breaking.

Based on the results, it can seem that all bones are beyond one. Therefore, it means the bones are in a safe region for standing on various incline platforms.

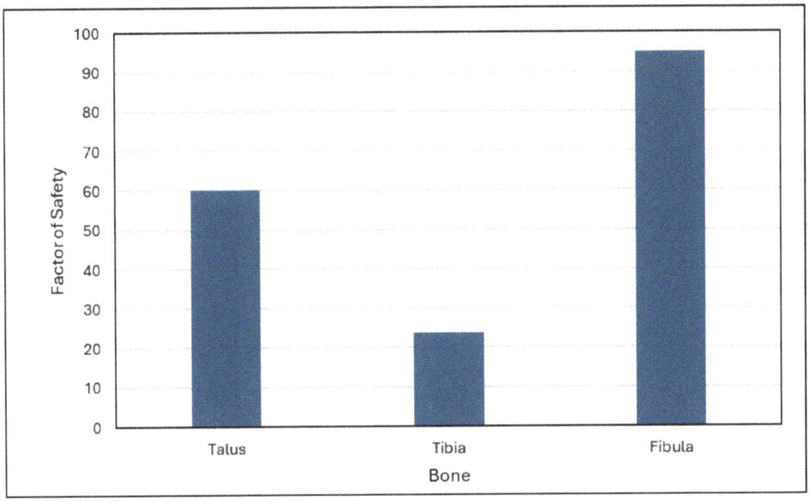

Figure 8. Factor of safety on ankle joint bones.

4. Discussion

The FE model developed in the present study was analysed based on von Mises stress responses at the ankle for slope standing. The analysis was performed under static linear elastic and the stresses obtained at the ankle part was assumed to not exceed the yield limit of the materials. Von Mises stress was used to evaluate the failure risk for the ankle joint because it considered the combined effect of normal and shear stresses. Based on the results, there are a few potential reasons that contribute to the increment of stress at the ankle joint component during incline and decline standing compared to flat standing.

Firstly, incline standing that involves a steeper slope requires the foot to be dorsiflexed, and similarly, declined standing requires the foot to be dorsiflexed to accommodate the slope [6]. Dorsiflexion is where the foot bends towards the shin. Inclined and declined standing changes the range of motion (ROM) at which joints operate [16]. The ankle joint experiences different angles of dorsiflexion and plantarflexion during slope standing compared to flat standing, impacting the distribution of forces and stress on the bones of the ankle. The decreased ROM of ankle dorsiflexion may lead to an increase in ground reaction force (GRF) and loading [17], which leads to an increment of stress on the ankle joint.

Furthermore, the higher force generated during incline standing to move against gravity and the higher impact force absorbed and distributed during declined standing [18] lead to a higher stress on the bones of the ankle during slope standing compared to flat standing. This higher stress due to the impact forces and joint loading may contribute to the risk of injury related to signs of fatigue, fractures, and pain at the patellofemoral [19].

In addition, inclined standing requires more effort from the muscles that plantarflex the ankle (push the foot down), including the calf muscles. This increased activity puts an additional load on the talus bone. As for declined standing, it requires the activation of eccentric muscles [16,19], especially in the lower leg, in order to control the descent. Eccentric muscle contractions involve lengthening of the muscle while producing force, and this can contribute to higher stress on the talus. As reported previously, the activation of the tibial muscle is changing due to the ankle response to the inclined surface during slope standing [20], and the magnitude and duration of activity in the gluteus maximus and hamstring are increased during incline standing [21,22]. Therefore, another possible reason that leads to stress increment on the bones of ankle joints is muscle engagement [16,23].

The combination of these factors results in an increased demand on the bones of the ankle during slope standing, leading to higher stress compared to standing on a flat surface. It is important to note that individuals with certain biomechanical issues, musculoskeletal conditions, or improper footwear may be more susceptible to increased stress and potential injury during slope standing. While slope standing can increase stress on the bones of the ankle joint, proper gait mechanics, muscle strength, and joint stability play crucial roles in mitigating excessive stress and preventing injuries.

In comparison to the maximum von Mises stress distribution view, the highest stress distribution was observed in ligaments compared to bones and cartilage. Ligaments, bones, and cartilage all undergo mechanical stresses during slope standing or any weight-bearing activity; but, because of their particular biomechanical characteristics and functions inside the body, the distribution and magnitude of these stresses might differ. The possible explanation for these differences is due to the material properties of the component itself. Ligaments are fibrous connective tissues that can resist tensile forces. The tensile strength of ligaments is high but the ability to withstand compressive forces is restricted [24]. Instead of that, bones are robust and strong, able to sustain compressive loads [25]. Although cartilage can resist compression, it is also capable of bearing some tensile loads [26] but to a lesser extent than ligaments. Due to these differences in material properties, ligaments tend to experience higher stress under tensile loading compared to bones and cartilage.

Other than that, specific anatomical functions might explain the various von Mises stress distributions found at each component of the ankle. Ligaments primarily function to stabilise joints by reducing excessive movement and providing structural support [27]. During slope standing, ligaments may uphold a significant portion of the load to maintain joint stability. Bones, on the other hand, support the body's weight and provide a framework for muscle attachment [28]. Cartilage acts as a cushion between bones and helps distribute forces within joints [29]. However, it generally experiences lower stresses compared to ligaments and bones. There are some limitations that are associated with this study. First, the cartilage and ligament of the ankle joint were assumed as linear elastic. Although these soft tissues can be considered as viscoelastic materials, the present analysis was performed under linear elastic region. The loads applied are relatively low compared to the material yield limit of all ankle parts. There are even studies that show that linear elastic model can stimulate bulk behaviour for cartilage accurately [30,31]. Despite limitations of the present model, particularly on ligament, the results showed that the stresses obtained on bones are not much different compared to other studies [11,15]. Hence, this limitation does not give a significant effect on the present findings. Second, the cartilage and ligaments were constructed to ensure the connectivity on foot structure that mainly consisted of bone. Other soft tissues such as skin and plantar fascia were omitted. Future research considering all soft tissues in the foot may need to be conducted to observe the effect of parts' interaction on the ankle biomechanics. Moreover, simulation of the ankle joint as a flexure hinge in the field of compliant mechanism is a quite interesting topic to be explored in future, particularly for prosthetic design.

5. Conclusions

This study aimed to develop finite element models of the ankle joint with various inclined foot postures and to analyse the stress distributions on the ankle joint due to an inclined or declined surface. This study successfully developed the finite element model of the ankle joint with various inclined foot postures. This study reported that, in general, the components of the ankle joint (bone, cartilage, and ligament) experienced different von Mises stress distribution patterns during flat standing in comparison with slope standing. In addition, this study found that the maximum von Mises stress distribution at the component of the ankle joint is higher during slope standing than flat standing. Taken together, these results suggest that slope standing, both inclined and declined more than 10 °, might contribute to a higher risk of injury as higher maximum stress was observed. Therefore, a better precaution and adaptation strategy should be taken during standing on slopes such as maintaining a proper body posture to ensure weight is evenly distributed at both feet, which can reduce stress at the ankle.

Author Contributions: Contributions of the authors are listed as follows: Conceptualisation, N.A.A.A.Y., K.S.B., M.F.I., M.H.M.S. and M.N.S.S.A.; Methodology, N.A.A.A.Y., K.S.B., M.F.I., M.H.M.S. and H.T.; Investigation, N.A.A.A.Y., K.S.B., M.F.I., M.N.S.S.A. and H.T.; Writing—original draft preparation, N.A.A.A.Y., K.S.B. and M.N.S.S.A.; Writing—review and editing, M.F.I., M.H.M.S. and H.T.; Funding acquisition, M.F.I. All authors have read and agreed to the published version of the manuscript.

Funding: The authors extend their appreciation to the Researchers Supporting Project (project number RSPD2024R1072), King Saud University, Riyadh, Saudi Arabia.

Institutional Review Board Statement: Not applicable.

Informed Consent Statement: Not applicable.

Data Availability Statement: The raw data supporting the conclusions of this article will be made available by the authors on request.

Acknowledgments: The authors extend their appreciation to the Researchers Supporting Project number (RSPD2024R1072), King Saud University, Riyadh, Saudi Arabia.

Conflicts of Interest: The authors declare no conflicts of interest.

References

1. Li, J.; Zhang, Y.; Song, S.; Hou, Y.; Hong, Y.; Yue, S.; Li, K. Dynamical Analysis of Standing Balance Control on Sloped Surfaces in Individuals with Lumbar Disc Herniation. *Sci. Rep.* **2020**, *10*, 1676. [CrossRef] [PubMed]
2. Dutt-Mazumder, A.; Gruben, K.G. Modulation of sagittal-plane center of pressure and force vector direction in human standing on sloped surfaces. *J. Biomech.* **2021**, *119*, 110288. [CrossRef] [PubMed]
3. Lee, J.W.; Park, J.S. The Correlation between Proprioception and Postural Control in Healthy Adults. *Iran. J. Public Health.* **2022**, *51*, 2360–2361. [CrossRef]
4. Park, S.Y.; Kim, S.H.; Park, D.J. Effect of slope angle on muscle activity during variations of the Nordic exercise. *J. Exerc. Rehabil.* **2019**, *15*, 832–838. [CrossRef]
5. Murray, K.J.; Azari, M.F. Leg length discrepancy and osteoarthritis in the knee, hip and lumbar spine. *J. Can. Chiropr. Assoc.* **2015**, *59*, 226–237. [PubMed]
6. Kwon, Y.; Shin, G. Foot kinematics and leg muscle activation patterns are altered in those with limited ankle dorsiflexion range of motion during incline walking. *Gait Posture* **2022**, *92*, 315–320. [CrossRef]
7. Taha, Z.; Norman, M.S.; Omar, S.F.S.; Suwarganda, E. A Finite Element Analysis of a Human Foot Model to Simulate Neutral Standing on Ground. *Procedia Eng.* **2016**, *147*, 240–245. [CrossRef]
8. Cheung, J.T.M.; Zhang, M.; Leung, A.K.L.; Fan, Y.B. Three-dimensional finite element analysis of the foot during standing—A material sensitivity study. *J. Biomech.* **2005**, *38*, 1045–1054. [CrossRef]
9. Park, S.; Lee, S.; Yoon, J.; Chae, S.W. Finite element analysis of knee and ankle joint during gait based on motion analysis. *Med. Eng. Phys.* **2019**, *63*, 33–41. [CrossRef]
10. Guan, M.; Zhao, J.; Kuang, Y.; Li, G.; Tan, J. Finite element analysis of the effect of sagittal angle on ankle joint stability in posterior malleolus fracture: A cohort study. *Int. J. Surg.* **2019**, *70*, 53–59. [CrossRef]
11. Ozen, M.; Sayman, O.; Havitcioglu, H. Modeling and stress analyses of a normal foot-ankle and a prosthetic foot-ankle complex. *Acta Bioeng. Biomech.* **2013**, *15*, 19–27. [PubMed]
12. Wang, Y.; Li, Z.; Wong, D.W.C.; Cheng, C.K.; Zhang, M. Finite element analysis of biomechanical effects of total ankle arthroplasty on the foot. *J. Orthop. Transl.* **2018**, *12*, 55–65. [CrossRef] [PubMed]
13. Hansen, A.H.; Childress, D.S.; Miff, S.C. Roll-over characteristics of human walking on inclined surfaces. *Hum. Mov. Sci.* **2004**, *23*, 807–821. [CrossRef] [PubMed]
14. Malakoutikhah, H.; de Cesar Netto, C.; Madenci, E.; Latt, L.D. Evaluation of assumptions in foot and ankle biomechanical models. *Clin. Biomech.* **2022**, *100*, 105807. [CrossRef]
15. Mondal, S.; Ghosh, R. A numerical study on stress distribution across the ankle joint: Effects of material distribution of bone, muscle force and ligaments. *J. Orthop.* **2017**, *14*, 329–335. [CrossRef]
16. Pickle, N.T.; Grabowski, A.M.; Auyang, A.G.; Silverman, A.K. The functional roles of muscles during sloped walking. *J. Biomech.* **2016**, *49*, 3244–3251. [CrossRef]
17. Almansoof, H.S.; Nuhmani, S.; Muaidi, Q. Role of ankle dorsiflexion in sports performance and injury risk: A narrative review. *Electron. J. Gen. Med.* **2023**, *20*, em521. [CrossRef]
18. Hong, Y.N.G.; Lee, J.; Shin, C.S. Transition versus continuous slope walking: Adaptation to change center of mass velocity in young men. *Appl. Bionics Biomech.* **2018**, 2018, 2028638. [CrossRef] [PubMed]
19. Park, S.K.; Jeon, H.M.; Lam, W.K.; Stefanyshyn, D.; Ryu, J. The effects of downhill slope on kinematics and kinetics of the lower extremity joints during running. *Gait Posture* **2019**, *68*, 181–186. [CrossRef]
20. Bavdek, R.; Zdolšek, A.; Strojnik, V.; Dolenec, A. Peroneal muscle activity during different types of walking. *J. Foot Ankle Res.* **2018**, *11*, 50. [CrossRef]

21. Wall-Scheffler, C.M.; Chumanov, E.; Steudel-Numbers, K.; Heiderscheit, B. Electromyography activity across gait and incline: The impact of muscular activity on human morphology. *Am. J. Phys. Anthr.* **2011**, *143*, 601–611. [CrossRef] [PubMed]
22. Franz, J.R.; Lyddon, N.E.; Kram, R. Mechanical work performed by the individual legs during uphill and downhill walking. *J. Biomech.* **2012**, *45*, 257–262. [CrossRef] [PubMed]
23. Liu, Y.; Qiang, L.; Song, Q.; Zhao, M.; Guan, X. Effects of backpack loads on leg muscle activation during slope walking. *Appl. Sci.* **2020**, *10*, 4890. [CrossRef]
24. Hoffmann, A.; Gross, G. Tendon and ligament engineering in the adult organism: Mesenchymal stem cells and gene-therapeutic approaches. *Int. Orthop.* **2007**, *31*, 791–797. [CrossRef] [PubMed]
25. Hart, N.H.; Nimphius, S.; Rantalainen, T.; Ireland, A.; Siafarikas, A.; Newton, R.U. Mechanical basis of bone strength: Influence of bone material, bone structure and muscle action. *J. Musculoskelet. Neuronal Interact.* **2017**, *17*, 114–139. [PubMed]
26. Mostakhdemin, M.; Nand, A.; Ramezani, M. Articular and artificial cartilage, characteristics, properties and testing approaches—A review. *Polymers* **2021**, *13*, 2000. [CrossRef]
27. Jung, H.-J.; Fisher, M.B.; Woo, S.L.-Y. Role of biomechanics in the understanding of normal, injured, and healing ligaments and tendons. *BMC Sports Sci. Med. Rehabil.* **2009**, *1*, 9. [CrossRef]
28. Su, N.; Yang, J.; Xie, Y.; Du, X.; Chen, H.; Zhou, H.; Chen, L. Bone function, dysfunction and its role in diseases including critical illness. *Int. J. Biol. Sci.* **2019**, *15*, 776–787. [CrossRef]
29. Fox, A.J.S.; Bedi, A.; Rodeo, S.A. The basic science of articular cartilage: Structure, composition, and function. *Sports Health.* **2009**, *1*, 461–468.
30. Zhang, K.; Li, L.; Yang, L.; Shi, J.; Zhu, L.; Liang, H.; Wang, X.; Yang, X.; Jiang, Q. Effect of degenerative and radial tears of the meniscus and resultant meniscectomy on the knee joint: A finite element analysis. *J. Orthop. Transl.* **2019**, *18*, 20–31. [CrossRef]
31. Bell, J.S.; Winlove, C.P.; Smith, C.W.; Dehghani, H. Modeling the steady-state deformation of the solid phase of articular cartilage. *Biomaterials* **2009**, *30*, 6394–6401. [CrossRef] [PubMed]

applied sciences

MDPI

Article

Does a Greater Hamstring Muscle Thickness Mean a Greater Aponeurosis Thickness?

Chrysostomos Sahinis and Eleftherios Kellis *

Laboratory of Neuromechanics, Department of Physical Education and Sport Sciences at Serres, Aristotle University of Thessaloniki, Agios Ioannis, 62110 Serres, Greece; sachinic@phed-sr.auth.gr

* Correspondence: ekellis@phed-sr.auth.gr; Tel.: +30-6988043758

Abstract: The aim of the present study was to assess the association between muscle thickness and the superficial and deep aponeurosis thickness of the hamstrings. Ultrasound images were captured from the semitendinosus (ST) and biceps femoris long head (BFlh) of fifty young individuals (28 males and 22 females). Measurements were taken at six sites along the thigh in prone position. Superficial and deep aponeurosis and muscle thickness were evaluated across the entire length of the ST and BFlh using a computational image segmentation approach, which generated approximately ~360 data points per participant. There was significant interindividual variability in superficial (coefficient of variation, CV: 19.37–31.40%) and deep aponeurosis thickness (CV: 21.22–31.82%). Correlation analysis revealed a limited association between muscle and superficial aponeurosis thickness, with no significant correlations ($r = -0.21$ to 0.22, $p > 0.05$) observed across regions. However, moderate positive correlations were found for deep aponeurosis thickness in the proximal region of the BFlh ($r = 0.43$, $p = 0.002$) and in the middle ($r = 0.34$, $p = 0.014$) and distal–middle region of the ST ($r = 0.35$, $p = 0.022$). No consistent relation between muscle and aponeurosis thickness was observed, suggesting that aponeurosis morphology varies independently of muscle size. This variability may impact the mechanical strain distribution, emphasizing the need for individualized assessments in hamstring injury prediction.

Keywords: hamstrings; ultrasound; aponeurosis; biceps femoris; muscle thickness

1. Introduction

Hamstring strain injuries (HSIs) are among the most frequent injuries in running-based and high-speed sports [1,2]. These injuries often lead to extended periods away from training and competition, diminishing athletic performance and causing lasting physical and structural changes [3,4]. The biceps femoris long head (BFlh) is susceptible to these injuries, often occurring at or near the musculotendinous junction (MTJ) during high-intensity eccentric contractions [1,2,5]. These contractions are especially prominent during the terminal swing phase of running gait, where rapid muscle lengthening under load places extreme mechanical stress on the muscle–tendon unit [1,2].

Aponeurosis morphology has a critical role in muscle function [6,7]. The aponeurosis is a connective tissue structure responsible for transmitting force from the muscle fibers to the tendon and ultimately to the skeletal system [6,7]. Its morphology (thickness, volume, and width) significantly influences the mechanical strain distribution within the muscle–tendon unit during contraction [6,7]. Abnormalities in these structural properties can lead to localized strain concentrations, particularly near the MTJ, which has been linked with greater risk of muscle injury [6,7]. Specifically, it has been hypothesized that disproportionate dimensions of the aponeurosis, such as a thinner or narrower aponeurosis, exacerbate strain concentrations, predisposing the muscle to injury during activities involving high mechanical loads [8,9]. Computational modeling and imaging-based assessments have indicated that strain maldistribution is most pronounced in regions adjacent to the MTJ [10–12]. Thus, understanding aponeurosis geometry is pivotal for identifying individuals at greater risk of HSIs.

The relationship between muscle and aponeurosis size is complex and characterized by significant variability [13–16]. Research has shown that the size of the vastus lateralis aponeurosis is directly related to the overall volume of the quadriceps [17]. In contrast, the size of the BFlh aponeurosis did not consistently correlate with muscle size, suggesting that aponeurosis dimensions are not proportionally scaled to the muscle's anatomical cross-sectional area or volume [13,15]. For instance, Evangelidis et al. [13] reported no relationship between aponeurosis area and knee flexor strength or muscle size, highlighting the high variability in aponeurosis-to-muscle area ratios. In contrast, Freitas et al. [14] reported moderate correlations ($r = 0.53$) between the BFlh muscle volume and BFlh aponeurosis volume. Additionally, Lazarczuk et al. [15] observed a significant positive moderate correlation between BFlh aponeurosis and muscle volume ($r = 0.64$) in a previously injured limb, but no correlation in uninjured or healthy limbs. These findings emphasize the complex relationship between muscle and aponeurosis dimensions, which is influenced by individual anatomical differences and by structural adaptations that may occur in response to mechanical loading or injury.

Investigations of aponeurosis dimensions have relied predominantly on magnetic resonance imaging (MRI) [13–16]. While MRI remains the gold standard for aponeurosis morphological assessments, its high cost and limited accessibility have constrained its application. However, recently, it has been found that ultrasound (US) is as a reliable and cost-effective alternative for evaluating aponeurosis thickness with relatively low measurement error (standard error of measurement 0.88 to 2.72%) [18].

While previous research has provided insights into the role of aponeurosis morphology in HSIs, several critical gaps persist. Prior studies have primarily focused on male populations [13–16], neglecting potential differences across genders. Furthermore, these investigations have been limited to assessing the BFlh, without considering other hamstring muscles such as the semitendinosus (ST), which may also play a pivotal role in injury risk and prevention. In addition, studies have shown significant variability in architecture along the hamstring muscle–tendon unit [19]. This means that the relationship, if any, between muscle and aponeurotic tissue thickness may vary depending on the region that is selected for analysis. Furthermore, the morphology of the muscle–tendon unit may vary between individuals [13,14], which might also have an impact on the relationship between aponeurotic and muscle tissue thickness. If the thickness of the muscle and aponeurosis tissues varies along the muscle–tendon unit and varies significantly between individuals, then this means that conclusions regarding the relationship between muscle and aponeurosis thickness may largely depend on the measurement method as well as the participant characteristics. Additionally, while some studies use mean values [14,15] for the assessment of aponeurosis dimensions, others [16] have focused on maximum values to identify local peaks. These different methods, combined with significant individual variability, complicate comparisons and interpretations of the relationship between muscle and aponeurosis thickness across the different studies. Identifying such relationships may assist in better understanding if and how muscle morphology correlates with aponeurosis morphology in healthy people and then using this information to examine abnormalities. Therefore, the aim of this study is to investigate the relation between muscle thickness and the thickness of both the superficial and deep aponeurosis layers of young males and females at different measurement locations across the thigh length. It was hypothesized that there is a significant association in thickness between muscle and aponeurosis tissues across the different measurement sites. Further, it was also hypothesized that the measurement method will significantly influence the observed relationship between muscle thickness and aponeurosis thickness. Finally, large variability between individuals in aponeurosis dimensions was expected.

2. Materials and Methods

2.1. Participants

The study comprised 50 participants, including 28 males (age: 20.3 ± 1.87 years; body mass: 84.5 ± 5.33 kg; height: 1.84 ± 0.06 m) and 22 females (age: 20.5 ± 1.72 years; body mass: 73.4 ± 4.14 kg; height: 1.67 ± 0.07 m). All participants were physically active but did not follow any structured sports or exercise program during the study period. Physical activity was evaluated using the short format of the International Physical Activity Questionnaire [20], with an average energy expenditure of 1766 ± 899 MET·min·week^{-1}. Prior to participation, written informed consent was obtained, and participants confirmed the absence of any prior lower limb injuries, including hamstring- or knee-related muscle or ligament injuries. To standardize conditions, participants refrained from intense physical activity for at least 24 h before their scheduled visits. Ethical approval was granted by the university's Institutional Review Board (ERC-001/2024), and the study adhered to the principles outlined in the Declaration of Helsinki.

2.2. Experimental Procedure

We assessed the variations in thickness of the superficial and deep aponeuroses, as well as the muscle thickness of the ST and BFlh, following methodologies from our recent research [18], which demonstrated high test–retest reliability with minimal measurement error. Participants' hip joints were stabilized in a neutral position with elastic Velcro straps to prevent movement during assessments. Knee angular position was monitored using a twin-axis goniometer (Model TSD 130B, Biopac Systems, Inc., Goleta, CA, USA), with 0° representing full knee extension. All US recordings were obtained unilaterally by a single examiner with over five years of expertise in musculoskeletal US.

2.3. Ultrasound Protocol

A linear-array probe (10 MHz frequency, 6 cm length) connected to a US device (SSD-3500, ALOKA, Tokyo, Japan) was used for imaging. The US settings were kept the same for everyone, with a depth of 80 mm, gain at 59 dB, a dynamic range of 65 dB, and a frequency of 10 MHz. To measure the femur length, we used a precise digital caliper (Model EC10, Tresna, Guilin Guanglu Measuring Instrument Co., Ltd., Guilin, China) that could measure up to 0.005 mm. This measurement was taken from the greater trochanter to the outer femoral condyle. US scans, both longitudinal and transverse, were used to map the path of each hamstring muscle. These scans identified the starting (proximal) and ending (distal) attachment points of the BFlh and ST, which were marked on the skin. The proximal attachment for all muscles was at the ischial tuberosity, while the distal attachment was at the fibular head for the BFlh and the pes anserinus for the ST [21–23]. To take into consideration the intramuscular differences of hamstrings muscles, femur length was divided into six measurement sites (15%, 30%, 45%, 60%, 75%, and 90%) (Figure 1), recognizing that the hamstring muscles exhibit intramuscular variations in both morphological [19] and activation patterns [24]. Axial lines were drawn on the skin with a marker, and the US probe was centered at each marked location.

To ensure consistent imaging, the US probe was positioned perpendicular to the skin and adjusted slightly to improve the clarity of the muscle aponeurosis [18]. Gentle, steady pressure was applied, along with a water-based gel, to avoid distorting the tissue and to enhance sound transmission. Participants were asked to stay relaxed during the procedure to avoid any interference with the imaging. Two investigators independently reviewed the images to confirm their quality, and any unclear images were retaken.

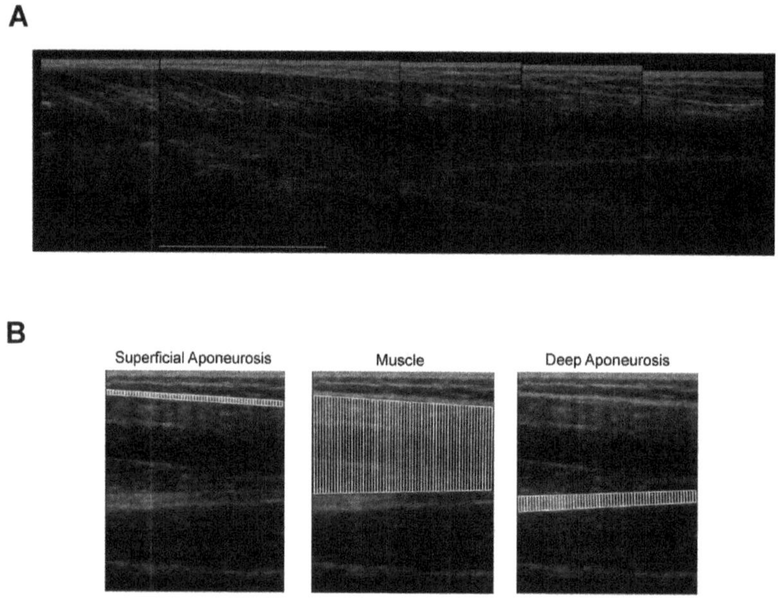

Figure 1. (**A**) Representative examples of a series of ultrasound images obtained at the different measurement sites along the thigh length for the biceps femoris. (**B**) Example of ultrasound images of biceps femoris muscle indicating region of interest as well as the placement of the landmarks to measure muscle and aponeurosis thickness. The fit provided set of landmarks, comprising x- and y-coordinates of 60 evenly spaced pixel points spanning the entire width of the region of interest. The white vertical lines represent thickness measures.

2.4. Image Analysis

A single-blinded examiner (C.S.) analyzed all US images using custom MATLAB scripts (MATLAB 2019b; MathWorks, Inc., Natick, MA, USA). Regions of interest (ROIs) included the superficial and deep aponeurosis and muscle thickness. Landmarks (x- and y-coordinates) were placed at 60 evenly spaced points along the width of each US image, resulting in ~360 data points per participant (6 images × 6 sites per muscle) [18]. Muscle thickness was measured as the distance between the superficial and deep aponeuroses (Figure 1). Aponeurosis thickness was determined as the distance between hyperechoic edges (Figure 1). Measurements were normalized to muscle length, enabling standardized comparisons between participants and muscle groups.

We used two different methods to examine the association between muscle thickness and the thickness of the superficial and deep aponeuroses. Firstly, we divided the muscle into five sections: proximal (15–30% of femur length, near to the hip joint), proximal–middle (30–45% of femur length), middle (45–60% of femur length), distal–middle (60–75% of femur length), and distal (75–90% of femur length, near to the knee joint). For each section, we used two approaches: (1) the first involved calculating the mean thickness of the aponeurosis and muscle within each region (averaging ~72 data points per segment), while (2) the second method focused on identifying the maximum thickness values within each region.

2.5. Statistical Analysis

Data were analyzed using SPSS Statistics for Windows Version 28.0 (Version 25.0, IBM, Armonk, NY, USA). The association between muscle thickness and aponeurosis thickness was investigated with multivariate regression analysis, adjusted for physical activity, body mass, and gender. Product interaction terms between aponeurosis thickness and these confounding variables were also considered in order to investigate the effects of moderators. The model estimated both main effects and interaction terms to test for whether the muscle–aponeurosis association differed by physical activity level, body mass, or gender. The fit of the overall model was determined with R^2, while the alpha level of significance was set at $p < 0.05$. Also, we performed correlation analyses using Pearson's or Spearman's correlation coefficients. Pearson's correlation was used for variables with a normal distribution, as determined by the Shapiro–Wilk test. In cases where the data were non-normally distributed, Spearman's rank-order correlation was applied. The strength of a correlation coefficient was interpreted in the following way: 0–0.10: trivial, 0.10–0.39: weak, 0.40–0.69: moderate, and 0.70–0.99: strong [25]. These analyses were conducted separately for the entire muscle length and for each of the five segmented regions (proximal, proximal–middle, middle, distal–middle, and distal). Additionally, the coefficient of variation (CV) was computed and expressed as a percentage to evaluate the variability in muscle and aponeurosis thickness of the BFlh and ST across the different measurement sites.

3. Results

3.1. Superficial Aponeurosis Thickness

3.1.1. Mean Thickness

The multivariate regression analysis for mean muscle thickness and superficial aponeurosis thickness present trivial to weak relationships for the BFlh (R^2 range = 0.11 to 0.18) and ST (R^2 range = 0.06 to 0.27) after adjusting for physical activity levels, body mass, and gender (Table 1). Additionally, the correlation analyses for the mean muscle thickness and the mean superficial aponeurosis thickness across the five measurement sites are presented in Figure 2. For the BF, none of the regions demonstrated significant correlations (r range = −0.21 to 0.22, $p > 0.05$). Similarly, correlations were not significant in any region (r range = 0.03 to 0.14, $p > 0.05$) for the ST across the different measurement sites.

Table 1. Regression analysis results examining the relationship between muscle thickness and deep aponeurosis thickness, controlling for the confounding factors physical activity, body mass, and gender ($n = 50$).

	Biceps Femoris		Semitendinosus	
	R^2	p	R^2	p
	Mean Thickness			
	Superficial Aponeurosis			
Proximal	0.18	0.27	0.27	0.06
Proximal–Middle	0.11	0.66	0.07	0.90
Middle	0.18	0.28	0.13	0.52
Distal–Middle	0.11	0.61	0.10	0.67
Distal	0.14	0.48	0.06	0.92
	Deep Aponeurosis			
Proximal	0.34	0.01	0.23	0.13
Proximal–Middle	0.20	0.18	0.08	0.86

Table 1. *Cont.*

	Biceps Femoris		Semitendinosus	
	R^2	*p*	R^2	*p*
Middle	0.14	0.45	0.10	0.67
Distal–Middle	0.20	0.19	0.20	0.18
Distal	0.16	0.35	0.07	0.93
Maximum Thickness				
Superficial Aponeurosis				
Proximal	0.37	0.01	0.11	0.61
Proximal–Middle	0.26	0.05	0.07	0.84
Middle	0.25	0.07	0.09	0.79
Distal–Middle	0.25	0.08	0.15	0.42
Distal	0.08	0.80	0.08	0.80
Deep Aponeurosis				
Proximal	0.30	0.03	0.14	0.46
Proximal–Middle	0.22	0.15	0.11	0.65
Middle	0.26	0.07	0.24	0.10
Distal–Middle	0.20	0.20	0.24	0.09
Distal	0.16	0.35	0.09	0.80

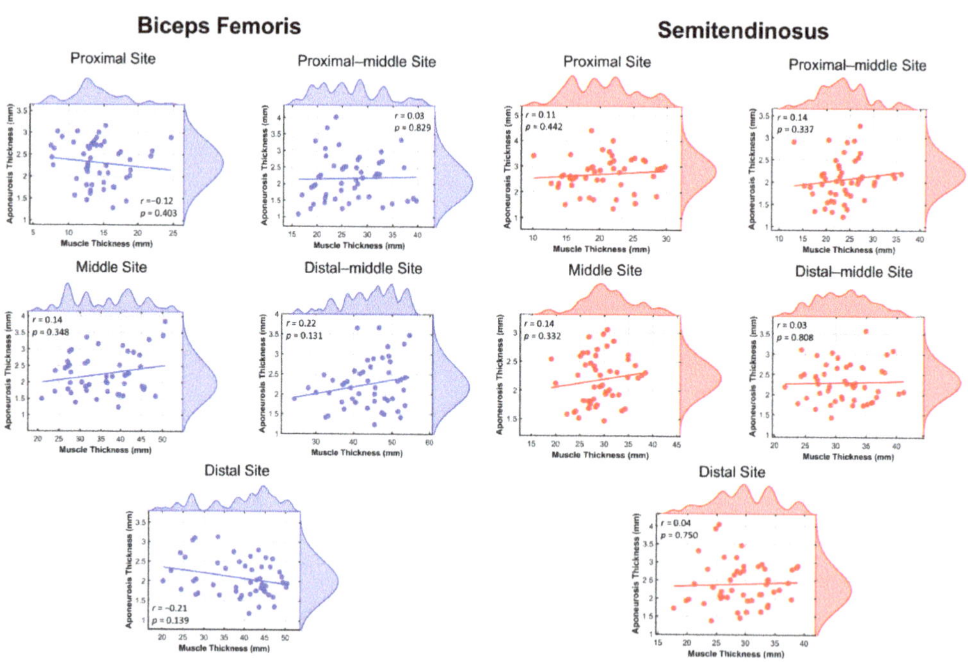

Figure 2. Scatterplots with density distributions showing the relationship between mean muscle thickness (x-axis, in mm) and mean superficial aponeurosis thickness (y-axis, in mm) across different measurement sites for the biceps femoris (blue) and semitendinosus (red) muscles. Measurement sites include proximal, proximal–middle, middle, distal–middle, and distal regions. Each point represents an individual observation (n = 50), with corresponding density plots indicating the data distribution at each site.

3.1.2. Maximum Thickness

The multivariate regression analysis for maximum muscle thickness and superficial aponeurosis thickness indicated trivial to weak relationships for the BFlh (R^2 range = 0.08 to 0.37) and for the ST (R^2 range = 0.07 to 0.15), adjusting for physical activity levels, body mass, and gender (Table 1). Additionally, the correlation analyses for the maximum muscle thickness and the maximum superficial aponeurosis thickness across the five measurement sites are presented in Figure 3. For the BF, the proximal site demonstrated a strong positive association, with a significant correlation (r = 0.47, $p < 0.001$, 95% CI [0.21, 0.66]). No significant correlations were observed for the remaining measurement sites (r range = 0.05 to 0.27, $p > 0.05$). Similarly, for the ST, no significant correlations were observed across all regions (r range = −0.20 to 0.13, $p > 0.05$).

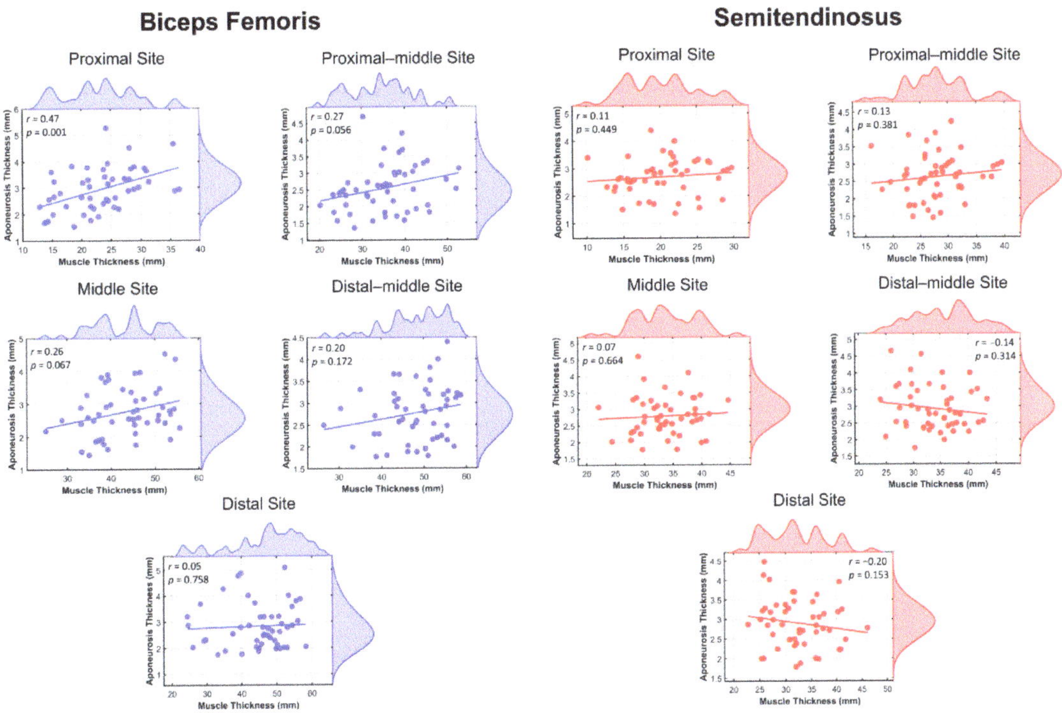

Figure 3. Scatterplots with density distributions showing the relationship between maximum muscle thickness (x-axis, in mm) and maximum superficial aponeurosis thickness (y-axis, in mm) across different measurement sites for the biceps femoris (blue) and semitendinosus (red) muscles. Measurement sites include proximal, proximal–middle, middle, distal–middle, and distal regions. Each point represents an individual observation (n = 50), with corresponding density plots indicating the data distribution at each site.

3.2. Deep Aponeurosis Thickness

3.2.1. Mean Thickness

The multivariate regression analysis for mean muscle thickness and deep aponeurosis thickness revealed trivial to weak relationships for the BFlh (R^2 range = 0.14 to 0.34) and for the ST (R^2 range = 0.07 to 0.24), after adjusting for confounding variables (Table 1). Figure 4 displays the correlation analyses for the mean muscle thickness and the mean thickness of the deep aponeurosis across the five measurement sites. For the BFlh, a significant positive moderate correlation was observed in the proximal region (r = 0.43, $p = 0.002$, 95% CI [0.17,

0.63]), while the correlations in the proximal–middle region through to the distal region were not significant (r range = 0.18 to 0.28, $p > 0.05$). For the ST, a significant positive weak correlation was identified in the distal–middle region (r = 0.28, $p = 0.049$, 95% CI [0.002, 0.52]). However, the correlations in the other measurement sites were non-significant (r range = −0.07 to 0.12, $p > 0.05$).

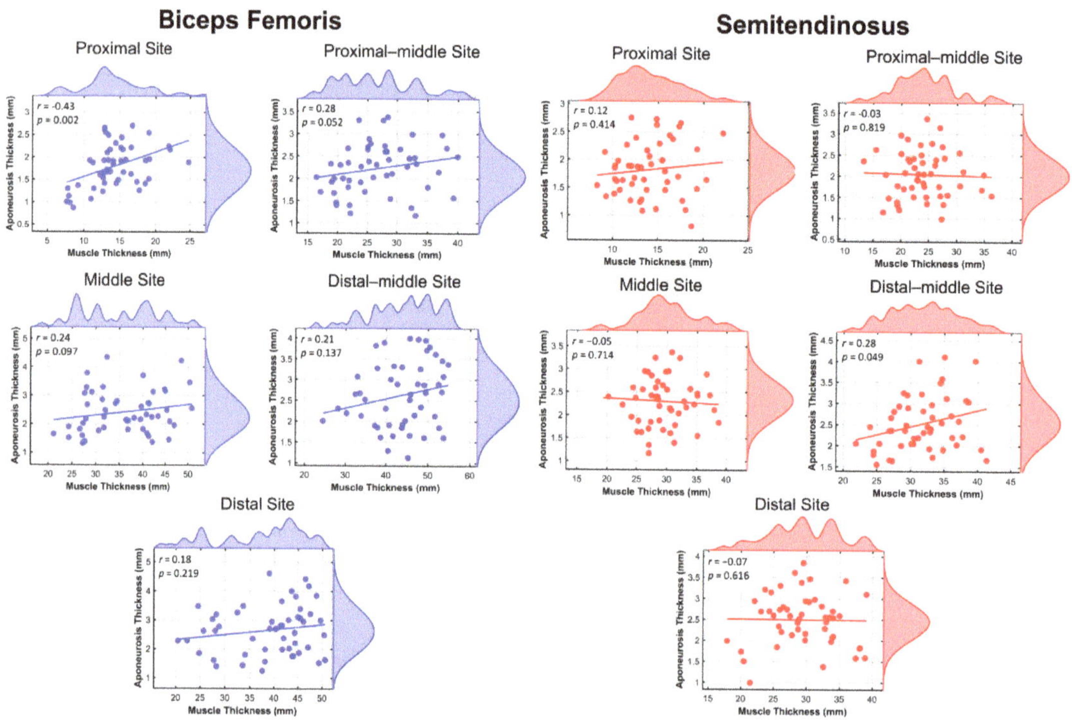

Figure 4. Scatterplots with density distributions showing the relationship between mean muscle thickness (x-axis, in mm) and mean deep aponeurosis thickness (y-axis, in mm) across different measurement sites for the biceps femoris (blue) and semitendinosus (red) muscles. Measurement sites include proximal, proximal–middle, middle, distal–middle, and distal regions. Each point represents an individual observation ($n = 50$), with corresponding density plots indicating the data distribution at each site.

3.2.2. Maximum Thickness

After adjusting for confounding variables, the multivariate regression analysis for maximum muscle thickness and deep aponeurosis thickness revealed trivial to weak associations for the BFlh (R^2 range = 0.16 to 0.30) and for the ST (R^2 range = 0.09 to 0.24). Figure 5 illustrates the correlation analyses between the maximum muscle thickness and the maximum thickness of the deep aponeurosis across the five measurement sites. For the BF, none of the regions showed significant correlations (r range = 0.13 to 0.27, $p > 0.05$). For the ST, significant positive correlations were identified in the middle site (r = 0.34, $p = 0.014$, 95% CI [0.073, 0.568]) and distal–middle site (r = 0.35, $p = 0.022$, 95% CI [0.049, 0.552]). The rest of the measurement sites did not exhibit significant correlations (r range = 0.001 to 0.12, $p > 0.05$).

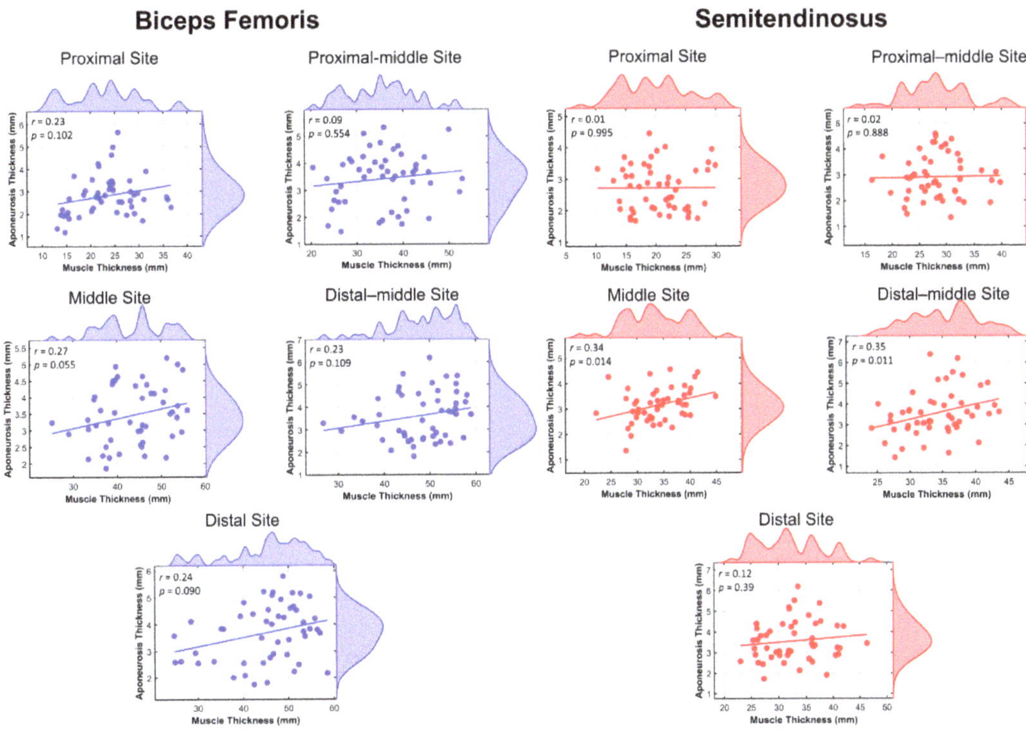

Figure 5. Scatterplots with density distributions showing the relationship between maximum muscle thickness (x-axis, in mm) and maximum deep aponeurosis thickness (y-axis, in mm) across different measurement sites for the biceps femoris (blue) and semitendinosus (red) muscles. Measurement sites include proximal, proximal–middle, middle, distal–middle, and distal regions. Each point represents an individual observation (n = 50), with corresponding density plots indicating the data distribution at each site.

3.2.3. Interindividual Variability in Muscle and Aponeurosis Thickness

The CV for muscle and superficial and deep aponeurosis thickness across the five measurement sites of the BFlh and ST are summarized in Table 2. For mean muscle thickness, the BFlh exhibited a CV range between 16.88% (distal–middle site) and 25.98% (proximal site), while the maximum thickness varied from 15.49% (distal–middle) to 25.90% (proximal site). In the ST, the CV for the mean thickness ranged from 14.53% (middle site) to 22.82% (proximal site), while the maximum thickness ranged from 13.97% (middle site) to 23.56% (proximal site).

Regarding superficial aponeurosis thickness, the BFlh showed a CV for mean thickness of 20.93% (proximal site) to 31.40% (proximal–middle site) and for maximum thickness, 22.71% (distal–middle site) to 28.00% (proximal–middle). In the ST, the CV for mean thickness ranged from 19.37% (middle site) to 24.88% (distal site), while the maximum thickness spanned 21.10% (middle site) to 28.90% (proximal–middle site).

For deep aponeurosis thickness, the BFlh demonstrated a CV for mean thickness of 24.75% (distal–middle site) to 31.52% (distal–middle site), and for maximum thickness, 24.75% (distal site) to 31.82% (proximal site). In the ST, the CV for mean thickness ranged from 23.22% (middle site) to 27.00% (proximal–middle site), while the maximum thickness ranged from 21.22% (middle site) to 31.15% (distal–middle site).

Table 2. Coefficient of variation (%) for the mean and maximum measurements of the muscle as well as the superficial and deep aponeurosis thickness of the biceps femoris long head (BFlh) and semitendinosus (ST) muscles across five measurement sites ($n = 50$).

Measurement Site	Muscle				Superficial Aponeurosis				Deep Aponeurosis			
	BFlh		ST		BFlh		ST		BFlh		ST	
	Mean	Max	Mean	Max	Mean	Max	Mean	Max	Mean	Max	Mean	Max
Proximal	25.98	25.90	22.82	23.56	20.93	27.70	23.33	24.24	24.75	31.82	26.32	26.90
Proximal–Middle	22.70	21.94	20.01	18.10	31.40	28.00	22.54	24.41	30.76	28.78	27.00	28.90
Middle	21.26	17.11	14.53	13.97	26.64	26.48	19.37	21.10	29.86	25.31	23.22	21.22
Distal–Middle	16.88	15.49	14.69	14.40	26.46	22.71	20.07	22.37	31.52	29.12	25.47	31.15
Distal	21.20	19.54	17.45	16.49	22.19	29.69	24.88	21.32	24.75	28.68	24.64	27.25

4. Discussion

This study examined the relationship between muscle thickness and the thickness of both the superficial and deep aponeuroses in the BFlh and ST muscles of young males and females. The results showed no consistent correlation between these parameters across the different measurement sites, suggesting that aponeurosis thickness may not be directly related to muscle thickness in the hamstring muscles even after adjusting for confounding variables such as physical activity, body weight, and gender. The considerable variability in aponeurosis thickness between individuals likely contributed to the low correlation coefficients. Additionally, the correlation remained relatively low even when mean thickness was used instead of maximum thickness.

4.1. No Association Between Muscle and Aponeurosis Thickness

Our results demonstrated limited correlations between muscle thickness and superficial aponeurosis thickness across different regions (Table 1, Figures 2 and 3). In contrast, stronger correlations were observed for the deep aponeurosis, particularly in the proximal region of the BFlh and the distal–middle region of the ST (Table 1, Figures 4 and 5). This challenges the common assumption as well as our initial hypothesis that aponeurosis and muscle morphology are proportionally aligned to accommodate force transmission during muscle contraction [7,8]. Previous studies have indicated that aponeurosis geometry influenced strain distribution during muscle loading. For example, Rehorn and Blemker [9], using computational modeling, demonstrated that thinner or narrower aponeuroses increase localized strain concentrations near the MTJ.

The absence of consistent correlations between muscle and aponeurosis size aligns with the findings of Evangelidis et al. [13], suggesting that the aponeurosis area in the BFlh is highly variable and does not correlate with muscle cross-sectional area or volume. Also, the present findings are in line with the results of Lazarczuk et al. [15], which reported no association between BFlh aponeurosis and muscle volume in uninjured or healthy limbs. However, the present findings are not in line with Freitas et al. [14], who found moderate correlations ($r = 0.53$) between the BFlh muscle volume and BFlh aponeurosis volume. In summary, these findings emphasize the complex relation between muscle and aponeurosis dimensions (thickness), highlighting the need for further research to better understand their functional and structural interactions.

4.2. Hamstrings' Morphological Characteristics Varied Across Individuals

The present results suggest significant interindividual (CV range 20.93–31–82%, Table 2) variability in aponeurosis thickness, as indicated by the density distribution plots. Similarly, Evangelidis et al. [13] reported up to a fourfold variability in the proximal BFlh aponeurosis dimensions across individuals, suggesting a high degree of heterogeneity in structural characteristics that could potentially influence hamstring mechanics. This high degree of variability (Table 2) may further influence the overall correlation values between muscle and aponeurosis thickness and could help explain the generally low correlations observed in this study. The present results extend the previous findings by demonstrating that both the superficial and deep aponeuroses exhibit high variability across young males and females at different regions of the hamstring muscles.

The variability observed in this study may reflect both intrinsic factors, such as muscle [19] and tendon architectural differences [21,22], and extrinsic factors, including previous injuries, training background, and adaptive remodeling [13–16]. For instance, Lazarczuk et al. [15] identified significant increases in aponeurosis volume in previously injured limbs, likely due to scar formation or altered loading patterns following injury. From a clinical perspective, this interindividual variability complicates the establishment of universal criteria for aponeurosis size in injury risk prediction. Instead, our findings advocate for personalized evaluations that consider the unique structural and functional characteristics of an individual's muscle–aponeurosis complex.

4.3. Mean vs. Maximum Thickness Assessment

This study's methodological approach provided both mean (Figures 2 and 4) and maximum thickness measurements (Figures 3 and 5) for the superficial and deep aponeuroses. While maximum thickness showed sporadic correlations with muscle thickness, mean values were generally consistent across regions between the two muscles. This discrepancy highlights the importance of capturing both mean and peak values to fully understand the structural characteristics of the superficial and deep aponeuroses. Maximum thickness could potentially reflect localized adaptations or stress concentrations that occur in response to region-specific loading patterns, whereas mean thickness could offer a broader representation of overall aponeurosis structure.

4.4. Practical Implications

The findings of this study have several practical implications. It is highly unlikely that an individual with thicker hamstring muscle tissue will also have thicker aponeurotic tissue, and vice versa, highlighting the importance of evaluating these tissues independently rather than assuming a proportional relationship. The large interindividual variability in the relationship between muscle and aponeurosis thickness suggests that the functional interactions between these tissues cannot be easily explained based on thickness alone. This complexity underscores the need for individualized evaluations, particularly following injury or intervention, as structural changes are specific to the individual being assessed. Clinicians should consider this variability when designing rehabilitation protocols or injury prevention strategies, as personalized assessments can provide better insights into tissue mechanics. From a research and diagnostic perspective, exploring additional factors beyond thickness, such as tissue stiffness or quality, may help to better understand muscle–aponeurosis interactions.

4.5. Limitations

While this study provides valuable insights, some limitations should be emphasized. This study focused on a physically active healthy population, which may limit the generalizability of the findings to elite or injured athletes. The selection of this population was used to attain homogeneity and reduce variations from different training regimens, inactivity, or previous injuries. This approach was essential to accurately focus on the relationship between muscle thickness and aponeurosis structure, providing baseline information for future comparisons with other groups. While US imaging is a reliable and accessible tool for assessing aponeurosis and muscle morphology [18], its two-dimensional nature and reliance on operator skill introduce limitations. Incorporating advanced imaging modalities, such as MRI or shear wave elastography, could provide further insights into the structural properties of aponeuroses.

5. Conclusions

This study did not identify significant associations between superficial and deep aponeurosis thickness and muscle thickness in the ST and BFlh muscles. However, substantial interindividual variability in aponeurosis thickness was observed. These findings highlight the need for personalized approaches in both research and clinical settings to account for the morphological diversity that may influence muscle–tendon unit function.

Author Contributions: Conceptualization, C.S. and E.K.; methodology, C.S. and E.K.; validation, C.S. and E.K.; formal analysis, C.S. and E.K.; data curation, C.S. and E.K.; writing—original draft preparation, C.S. and E.K.; writing—review and editing, C.S. and E.K.; visualization, C.S.; supervision, E.K. All authors have read and agreed to the published version of the manuscript.

Funding: This research received no external funding.

Institutional Review Board Statement: The protocol was approved by the Institutional Ethics Committee (ERC-001/2024).

Informed Consent Statement: Informed consent was obtained from all subjects involved in the study.

Data Availability Statement: The data produced in this paper are available at Sahinis, Chrysostomos; Kellis, Eleftherios (2024), "Does a greater hamstring muscle thickness mean a greater aponeurosis thickness?", Mendeley Data, V1, https://doi.org/10.17632/vnmhtktn8d.1.

Acknowledgments: We thank the participants in the study for volunteering their time.

Conflicts of Interest: The authors declare no conflicts of interest.

References

1. Askling, C.; Saartok, T.; Thorstensson, A. Type of acute hamstring strain affects flexibility, strength, and time to return to pre-injury level. *Br. J. Sports Med.* **2006**, *40*, 40–44. [CrossRef] [PubMed]
2. Koulouris, G.; Connell, D. Evaluation of the hamstring muscle complex following acute injury. *Skelet. Radiol.* **2003**, *32*, 582–589. [CrossRef] [PubMed]
3. Kellis, E.; Sahinis, C. Is Muscle Architecture Different in Athletes with a Previous Hamstring Strain ? A Systematic Review and Meta-Analysis. *J. Funct. Morphol. Kinesiol.* **2022**, *7*, 16. [CrossRef] [PubMed]
4. Timmins, R.G.; Shield, A.J.; Williams, M.D.; Lorenzen, C.; Opar, D.A. Architectural adaptations of muscle to training and injury: A narrative review outlining the contributions by fascicle length, pennation angle and muscle thickness. *Br. J. Sports Med.* **2016**, *50*, 1467–1472. [CrossRef]
5. Clanton, T.O.; Coupe, K.J. Hamstring strains in athletes: Diagnosis and treatment. *JAAOS J. Am. Acad. Orthop. Surg.* **1998**, *6*, 237–248. [CrossRef]
6. Azizi, E.; Roberts, T.J. Biaxial strain and variable stiffness in aponeuroses. *J. Physiol.* **2009**, *587*, 4309–4318. [CrossRef]
7. Lemos, R.R.; Epstein, M.; Herzog, W. Modeling of skeletal muscle: The influence of tendon and aponeuroses compliance on the force–length relationship. *Med. Biol. Eng. Comput.* **2008**, *46*, 23–32. [CrossRef]
8. Arellano, C.J.; Gidmark, N.J.; Konow, N.; Azizi, E.; Roberts, T.J. Determinants of aponeurosis shape change during muscle contraction. *J. Biomech.* **2016**, *49*, 1812–1817. [CrossRef]
9. Herzog, W. Skeletal muscle mechanics: Questions, problems and possible solutions. *J. Neuroeng. Rehabil.* **2017**, *14*, 98. [CrossRef]
10. Rehorn, M.R.; Blemker, S.S. The effects of aponeurosis geometry on strain injury susceptibility explored with a 3D muscle model. *J. Biomech.* **2010**, *43*, 2574–2581. [CrossRef]
11. Fiorentino, N.M.; Blemker, S.S. Musculotendon variability influences tissue strains experienced by the biceps femoris long head muscle during high-speed running. *J. Biomech.* **2014**, *47*, 3325–3333. [CrossRef] [PubMed]
12. Fiorentino, N.M.; Rehorn, M.R.; Chumanov, E.S.; Thelen, D.G.; Blemker, S.S. Computational models predict larger muscle tissue strains at faster sprinting speeds. *Med. Sci. Sports Exerc.* **2014**, *46*, 776–786. [CrossRef] [PubMed]

13. Evangelidis, P.E.; Massey, G.J.; Pain, M.T.G.; Folland, J.P. Biceps femoris aponeurosis size: A potential risk factor for strain injury? *Med. Sci. Sports Exerc.* **2015**, *47*, 1383–1389. [CrossRef]
14. Freitas, S.R.; Abrantes, F.; Santos, F.; Mascarenhas, V.; Oliveira, R.; Firmino, T.; Mendes, B.; Cerda, M.; Vaz, J.R. Is Biceps Femoris Aponeurosis Size an Independent Risk Factor for Strain Injury? *Int. J. Sports Med.* **2020**, *41*, 552–557. [CrossRef]
15. Lazarczuk, S.L.; Collings, T.J.; Hams, A.H.; Timmins, R.G.; Opar, D.A.; Edwards, S.; Shield, A.J.; Barrett, R.S.; Bourne, M.N. Biceps femoris long head muscle and aponeurosis geometry in males with and without a history of hamstring strain injury. *Scand. J. Med. Sci. Sports* **2024**, *34*, e14619. [CrossRef]
16. Balshaw, T.G.; McDermott, E.J.; Massey, G.J.; Hartley, C.; Kong, P.W.; Maden-Wilkinson, T.; Folland, J. Smaller biceps femoris aponeurosis size in legs with a history of hamstring strain injury. *Int. J. Sports Med.* **2024**, *45*, 871–883. [CrossRef]
17. Abe, T.; Kumagai, K.; Bemben, M.G. Muscle aponeurosis area in hypertrophied and normal muscle. *J. Trainol.* **2012**, *1*, 23–27. [CrossRef]
18. Sahinis, C.; Kellis, E. Reliability of Semitendinosus and Biceps Femoris Aponeurosis Thickness Using B-Mode Ultrasound. *J. Ultrasound Med.* **2024**, *9999*, 1–14. [CrossRef]
19. Kellis, E. Intra- and inter-muscular variations in hamstring architecture and mechanics and their implications for injury: A narrative review. *Sports Med.* **2018**, *48*, 2271–2283. [CrossRef]
20. Craig, C.L.; Marshall, A.L.; Sjöström, M.; Bauman, A.E.; Booth, M.L.; Ainsworth, B.E.; Pratt, M.; Ekelund, U.L.F.; Yngve, A.; Sallis, J.F. International physical activity questionnaire: 12-country reliability and validity. *Med. Sci. Sports Exerc.* **2003**, *35*, 1381–1395. [CrossRef]
21. Kellis, E.; Sahinis, C. Effect of knee joint angle on individual hamstrings morphology quantified using free-hand 3D ultrasonography. *J. Electromyogr. Kinesiol.* **2022**, *62*, 102619. [CrossRef] [PubMed]
22. Sahinis, C.; Kellis, E. Distal hamstrings tendons mechanical properties at rest and contraction using free-hand 3-D ultrasonography. *Scand. J. Med. Sci. Sports* **2024**, *34*, e14621. [CrossRef] [PubMed]
23. Woodley, S.J.; Mercer, S.R. Hamstring muscles: Architecture and innervation. *Cells Tissues Organs* **2005**, *179*, 125–141. [CrossRef] [PubMed]
24. Sahinis, C.; Amiridis, I.G.; Kannas, T.; Farina, D.; Enoka, R.M.; Kellis, E. Distinct neural drives along the semitendinosus muscle. *Med. Sci. Sports Exerc.* **2024**, *56*, 2338–2348. [CrossRef] [PubMed]
25. Schober, P.; Boer, C.; Schwarte, L.A. Correlation Coefficients: Appropriate Use and Interpretation. *Anesth. Analg.* **2018**, *126*, 1763–1768. [CrossRef]

applied sciences

MDPI

Article

Application of Mathematical Models for Blood Flow in Aorta and Right Coronary Artery

Monica Minniti †, **Vera Gramigna** *,†, **Arrigo Palumbo and Gionata Fragomeni**

Department of Medical and Surgical Sciences, University "Magna Graecia", Viale Europa, 88100 Catanzaro, Italy; monica.minniti@studenti.unicz.it (M.M.); a.palumbo@dimes.unical.it (A.P.); fragomeni@unicz.it (G.F.)
* Correspondence: gramigna@unicz.it
† These authors contributed equally to this work.

Abstract: Cardiovascular diseases represent one of the leading causes of mortality worldwide, underscoring the need for accurate simulations of blood flow to improve diagnosis and treatment. This study examines blood flow dynamics in two different vascular structures—the aorta and the right coronary artery (RCA)—using Computational Fluid Dynamics (CFD). Utilizing COMSOL Multiphysics®, various mathematical models were applied to simulate blood flow under physiological conditions, assuming a steady-flow regime. These models include both Newtonian and non-Newtonian approaches, such as the Carreau and Casson models, as well as viscoelastic frameworks like Oldroyd-B, Giesekus, and FENE-P. Key metrics—such as velocity fields, pressure distributions, and error analysis—were evaluated to determine which model most accurately describes hemodynamic behavior in large vessels like the aorta and in smaller and more complex vessels like the RCA. The results highlight the importance of shear-thinning and viscoelastic properties in small vessels like the RCA, which contrasts with the predominantly Newtonian behavior observed in the aorta. While computational challenges remain, this study contributes to a deeper understanding of blood rheology, enhancing the accuracy of cardiovascular simulations and offering valuable insights for diagnosing and managing vascular diseases.

Keywords: blood rheology; computational fluid dynamics (CFD); non-Newtonian models; viscoelasticity; shear-thinning models; cardiovascular simulation; right coronary artery (RCA); aorta hemodynamics

1. Introduction

Understanding the dynamics of blood flow in the human cardiovascular system is essential for advancing medical research and improving the diagnosis and treatment of vascular diseases. Blood, a complex non-Newtonian fluid, exhibits unique rheological behaviors that significantly influence its movement through vessels of varying sizes and geometries. This is particularly important in large arteries like the aorta and smaller vessels such as the right coronary artery (RCA), where variations in vessel structure and blood properties play a critical role in health outcomes. Computational fluid dynamics (CFD) has become a powerful tool to simulate these flows, offering detailed insights into hemodynamic phenomena that are difficult to capture through experimental methods alone.

The current state of research emphasizes the significance of accurately modeling blood flow, as it directly relates to the understanding of various cardiovascular diseases, such as atherosclerosis, aneurysms, and coronary artery disease. Previous studies [1–10] have applied both Newtonian and non-Newtonian fluid models to simulate blood flow, with non-Newtonian models—such as shear-thinning (Carreau, Casson) and viscoelastic

https://doi.org/10.3390/app15105748

frameworks (Oldroyd-B, Giesekus, FENE-P)—offering more accurate representations of real physiological conditions. While these models provide improved insights, they also come with challenges in terms of computational cost and complexity. Moreover, controversies persist regarding the optimal model choice, especially in regions where large variations in shear rates and vessel geometries exist.

While previous studies [1–10] have contributed significantly to the understanding of blood rheology and its role in hemodynamics, many present specific limitations. Some works focus primarily on theoretical models [2,4,8], lacking direct application to real vascular geometries. Others investigate isolated anatomical regions or pathological conditions without systematically comparing multiple rheological models [3,5,7]. Additionally, studies that explore viscoelastic properties often do so in vitro [9] or within simplified frameworks such as the Windkessel model [10], without evaluating spatial flow dynamics in anatomically accurate structures like the aorta and RCA. These gaps underscore the need for an integrative CFD analysis that compares both Newtonian and non-Newtonian models—including viscoelastic ones—across different vascular scales and geometries.

This study aims to investigate blood flow in the aorta and the RCA using COMSOL Multiphysics®, comparing Newtonian and non-Newtonian models to evaluate their effectiveness in replicating physiological conditions. By analyzing differences in flow behaviors, such as velocity profiles, pressure distributions, and shear stress dynamics, the research seeks to improve understanding of how blood interacts with complex vascular environments. These findings have the potential to inform the identification of risk factors for vascular diseases, and contribute to the development of more accurate diagnostic tools and personalized therapeutic strategies.

In conclusion, this study underscores the importance of selecting the appropriate model in CFD simulations to achieve accurate predictions of blood flow. It highlights the need for continued exploration in the field to refine our understanding of hemodynamics and support advancements in cardiovascular healthcare.

2. Materials and Methods

This study employed a range of mathematical models to describe the rheological behavior of blood, focusing on both Newtonian and non-Newtonian properties. These models were selected to represent the specific characteristics of blood flow under varying conditions, and each was implemented with sufficient detail to allow replication by others in the field.

2.1. Mathematical Models

1. Newtonian Model: Assumes a constant viscosity irrespective of the shear rate. This model is suitable for large-caliber vessels like the aorta, where shear rates are high, and blood behaves nearly like a Newtonian fluid [2–4]. The Newtonian model simplifies calculations, making it computationally efficient but less accurate for small or highly deformable vessels.
2. Non-Newtonian Shear-Thinning Models:
    - Carreau Model: An empirical model that describes the decrease in viscosity with increasing shear rates. It captures the Newtonian plateaus at low and high shear rates and the intermediate pseudoplastic region. The Carreau model is expressed as follows:

$$\mu(\dot{\gamma}) = \mu_\infty + (\mu_0 - \mu_\infty)\left[1 + (\lambda\dot{\gamma})^2\right]^{\frac{n-1}{2}}$$

where μ_∞: viscosity at infinite shear rate (in Pa · s), μ_0: viscosity at zero shear rate (in Pa · s), λ: relaxation time of the fluid (in seconds, s), $\dot{\gamma}$: shear rate (in s^{-1}), and n: power-law index (dimensionless), which determines the degree of shear-thinning behavior. For $n = 1$, the fluid behaves as a Newtonian fluid.
The fluid behavior depends on the value of $\dot{\gamma}$ relative to $\frac{1}{\lambda}$:

- If $\dot{\gamma} \ll \frac{1}{\lambda}$, the fluid behaves as a Newtonian fluid.
- If $\dot{\gamma} \gg \frac{1}{\lambda}$, the fluid exhibits power-law fluid behavior.

The parameter values of the Carreau Model can be considered as follows (as reported in Fernandes et al. [6]):

$$\mu_\infty = 0.0035\,\mathrm{Pa \cdot s}, \quad \mu_0 = 0.056\,\mathrm{Pa \cdot s}, \quad \lambda = 3.313\,\mathrm{s}, \quad n = 0.3568.$$

- Casson Model: Incorporates a yield stress (τ_y) below which the fluid does not flow. It is particularly suited for blood, considering the aggregation of red blood cells at low shear rates. Additionally, it predicts viscosity changes due to hematocrit (Hct) variations.
  The model is described by the following equations:

$$\mu = \frac{\tau_y}{\dot{\gamma}} + \frac{2\sqrt{\tau_y}\,N_\infty}{\sqrt{\dot{\gamma}}} + N_\infty^2$$

$$N_\infty = \frac{\sqrt{\mu_p}}{8\sqrt{1 - \mathrm{Hct}}}$$

  where μ: dynamic viscosity (in Pa · s), τ_y: yield stress, the minimum stress required for the fluid to flow (in Pa), $\dot{\gamma}$: shear rate (in s^{-1}), N_∞: a parameter related to viscosity changes due to hematocrit variations (dimensionless), μ_p: plasma viscosity, a constant property of the plasma (in Pa · s), and Hct: hematocrit, the volume fraction of red blood cells in the blood (dimensionless, typically expressed as a percentage).
  The parameter values for this model, as reported in Fernandes et al. [6], are as follows:

$$\tau_y = 0.625\,\mathrm{Hct}, \quad \mathrm{Hct} = 0.4, \quad \mu_p = 0.00145\,\mathrm{Pa \cdot s}.$$

  Here, τ_y is linearly proportional to Hct and the Hct value is 40% of blood volume. This model effectively captures the non-Newtonian behavior of blood and accounts for the effects of both yield stress and hematocrit on viscosity.

3. Viscoelastic Models [6]:
   - Oldroyd-B Model: The model describes blood as a viscoelastic fluid, combining a Newtonian solvent with elastic polymer-like components. Despite its computational simplicity, it lacks precision in capturing shear-thinning effects. The governing equation is given as follows:

$$\tau + \lambda_1 \left(\frac{\partial \tau}{\partial t} \right) = \eta_0 \left(\dot{\gamma} + \lambda_2 \frac{\partial \dot{\gamma}}{\partial t} \right)$$

     where τ represents the stress tensor, λ_1 is the relaxation time, λ_2 is the retardation time, η_0 is the zero-shear viscosity, and $\dot{\gamma}$ denotes the shear rate. This model provides a basic framework for studying viscoelastic behaviors in complex fluids such as blood.

- Giesekus Model: A nonlinear extension of the Oldroyd-B model, the Giesekus model incorporates a mobility parameter (α) to account for shear-thinning and elastic behaviors, making it more realistic for describing blood flow in geometrically complex vessels. The governing equation is expressed as follows:

$$\tau + \lambda_1 \left(\frac{\partial \tau}{\partial t} - \mathbf{L} \cdot \tau - \tau \cdot \mathbf{L}^T \right) + \frac{\alpha \lambda_1}{\eta_0} \tau \cdot \tau = \eta_0 \dot{\gamma},$$

  where α is the mobility parameter ($0 \leq \alpha \leq 1$), $\mathbf{L}$ represents the velocity gradient, τ is the stress tensor, λ_1 is the relaxation time, η_0 is the zero-shear viscosity, and $\dot{\gamma}$ denotes the shear rate.
- FENE-P Model (Finitely Extensible Nonlinear Elastic—Peterlin): Based on molecular theory, it describes the elasticity of long polymer chains while incorporating a finite extensibility factor. This makes it particularly suitable for modeling high-stress conditions, such as blood flow through stenosed arteries or within mechanical circulatory devices. The governing equation is expressed as follows:

$$\tau + \lambda \left(\frac{\partial \tau}{\partial t} - \mathbf{L} \cdot \tau - \tau \cdot \mathbf{L}^T \right) = \eta_0 \dot{\gamma}(f) + \frac{\lambda \eta_0}{\eta_p} (f \tau),$$

  where f is the finite extensibility factor (set to 100), η_p is the polymer viscosity, η_0 is the zero-shear viscosity (set to 0.03 Pa·s), and λ is the relaxation time (set to 0.001 s). The FENE-P model captures the nonlinear elastic behavior of polymers, making it highly effective for applications involving significant stress and deformation.

These models were implemented in COMSOL Multiphysics® (Version 6.3) software [11], allowing for detailed simulations of blood flow in both large (aorta) and small (right coronary artery, RCA) vessels. The ability of each model to replicate shear stress, velocity profiles, and pressure distributions was systematically analyzed, highlighting their respective strengths and limitations.

2.2. Geometric Models

The two vessels considered in our study are very different from each other in terms of size: the aorta, being the main vessel of the human body, has a diameter that typically varies between 20 and 30 mm, while the right coronary artery has a diameter significantly smaller, usually between 2 and 5 mm. This difference in size reflects their different functions: the aorta is responsible for distributing oxygenated blood throughout the body, while the RCA is responsible for supplying blood to the heart muscle. Finally, the geometry and flexibility of the coronary artery may vary (both inter- and intra-patient) more pronouncedly than that of the aorta. The coronary arteries therefore have a smaller diameter, more pronounced curvature compared to the aorta, and greater anatomical variability, factors which could give rise to shear-thinning behavior (reduction in viscosity with increasing shear rate) and other typical effects of non-Newtonian fluids. In contrast, the aorta, being a vessel of large diameter and having a more laminar flow, tends to behave like a Newtonian fluid, whose viscosity remains constant and proportional to the shear stress. This difference in rheological behavior between coronary arteries and the aorta is crucial for understanding blood flow dynamics and the clinical implications associated with cardiovascular disease.

- **Three-Dimensional Geometric Model of the Aorta**:

The geometric model used, developed by Caruso et al. [12], represents a specific aorta of a 50-year-old woman (height 160 cm), obtained from a CT scan performed for clinical purposes. The DICOM image sequence included 512 slices in the axial plane, 416 in the sagittal plane, and 88 in the coronal plane, with an in-plane resolution of 0.9375 × 0.9375 mm and a slice thickness of 1.4 mm. Written patient consent was obtained for the use of the medical images in the study, and this patient was chosen because she did not present any aortic diseases, such as aneurysms or dissections. The aorta was reconstructed through a segmentation process using the open-source software 3D Slicer. Subsequently, a reverse engineering algorithm was applied to obtain a more accurate virtual 3D model, suitable for computational analysis. The final model (shown in Figure 1) includes the ascending aorta, the aortic arch, the major vessels, and the descending trunk with abdominal vessels, and extends to the iliac arteries. The diameter of the aorta in the descending trunk is approximately 21 mm at the level of the aortic arch and 16 mm at the level of the renal arteries.

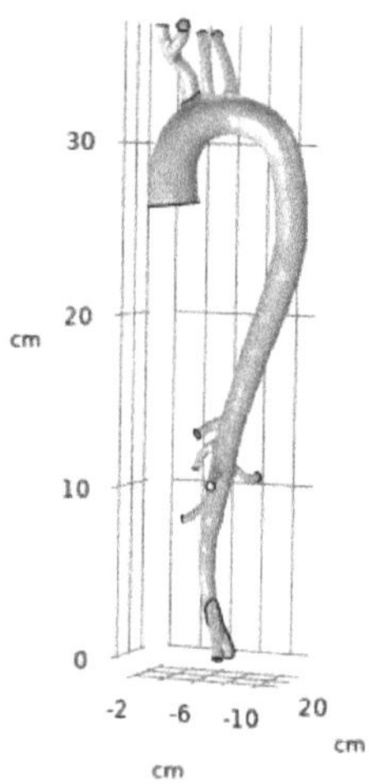

Figure 1. Aorta - 3D geometric model.

- **Three-Dimensional Geometric Model of the RCA**:

The 3D model of the right coronary artery, developed by Caruso et al. [13], is specific to a 64-year-old male patient and was obtained using a series of angiograms acquired during a standard angiography performed for clinical purposes. Written consent was obtained for the use of the medical images. Since the coronary arteries are compressed by the contraction of the cardiac muscle during systole, myocardial perfusion primarily occurs during the diastolic phase, when the muscle relaxes. For this reason, the geometric model was reconstructed considering the end of diastole. The true curvature of the vessel was obtained by combining the centerline path from the LAO (Left Anterior Oblique) and RAO (Right Anterior Oblique) views. The variation in diameter was modeled assuming a circular shape. Moreover, in order to obtain the real anatomy of the vessel, the known caliber of the catheter (6 Fr) was used, so the others measurements were obtained by means of a proportion. The RCA model (shown in Figure 2) includes the proximal segment, the mid-segment, and the distal segment (collateral branches were neglected).

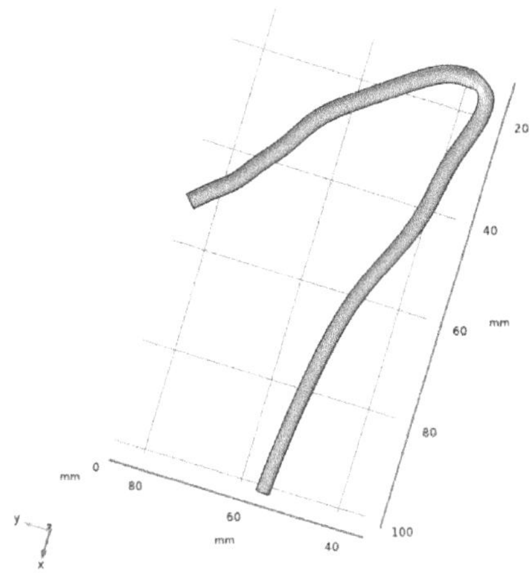

Figure 2. RCA - 3D geometric model.

2.3. Boundary Conditions

We defined the boundary conditions that, together with the initial conditions set to define the starting state of the fluid (blood) in the computational domain, were fundamental to ensure that the simulation was well posed and could proceed with the iterative process for solving the system of equations until convergence was reached. A no-slip condition was imposed on the walls, meaning zero velocity at these surfaces [14,15]. The rigid wall assumption causes small errors in the estimation of the fluid dynamic parameters. However, this should not influence our results, because the general flow characteristics remain similar [16,17]. Furthermore, the level of uncertainty in the patient-specific domain for vascular tissue, as well as the significant increase in computational costs of fluid–structure interaction (FSI) simulations, make this approximation currently the best compromise between accuracy and efficiency for clinical purposes [18–20]. In some cases, the stiffness of the vessels is increased so significantly that the hypothesis of a rigid wall may be valid. For steady-state simulations, the maximum value of the pulsatile flow rate (corresponding to the systolic peak) was applied at the inlet, while pressure was specified at the outlet. Specifically:

- Aorta: At $t = 0.13\,\mathrm{s}$, the inlet flow rate is $Q_{\mathrm{in}} = 450\,\mathrm{mL/s} = 450 \times 10^{-6}\,\mathrm{m^3/s}$ [21], and the outlet pressure is $P_{\mathrm{out}} = 70\,\mathrm{mmHg} = 9332.54\,\mathrm{Pa}$ [22].
- RCA: At $t = 0.2\,\mathrm{s}$, the inlet flow rate is $Q_{\mathrm{in}} = 2.25\,\mathrm{mL/s} = 2.25 \times 10^{-6}\,\mathrm{m^3/s}$ [23], and the outlet pressure is $P_{\mathrm{out}} = 133\,\mathrm{mmHg} = 17{,}731.83\,\mathrm{Pa}$ [24].

2.4. Simulation Details

In this study, COMSOL Multiphysics® 6.3 was used to simulate blood flow dynamics in the aorta and RCA. To discretize the computational domain, tetrahedral elements were employed, automatically generated by the software's default settings. Mesh quality was visually inspected to ensure the absence of distorted cells, which could compromise solution reliability. Both for the aorta and for the RCA, the mesh consisted of approximately 450k elements. Refinement was increased near walls, outlets, and geometric transitions to improve accuracy.

For the viscoelastic FENE-P model in the aorta, additional refinement was applied below the renal artery to address issues in this critical area, using the "Size" tool to enhance mesh density. The final simulations used a physics-controlled "fine" mesh size, balancing accuracy and computational efficiency. The solvers employed were GMRES for the aorta and both PARDISO and GMRES for the RCA, with convergence achieved at a residual error tolerance of 10^{-6}. The computational times required for solving the Navier–Stokes equations varied across models: aorta simulations took at least twice as long as RCA simulations and were even more computationally intensive for viscoelastic models.

To solve the Navier–Stokes equations, the following choices were made: the Pardiso solver with a pivoting perturbation of 1.0×10^{-13}, the P1–P1 finite element method for spatial discretization with linear elements for both velocity components and the pressure field, and the P1–P2 finite element method for viscoelastic simulations to ensure greater accuracy and stability in capturing complex flow behaviors.

2.5. Algorithm for Calculating the Error

Once the geometrical models were imported into COMSOL, various measurements were performed, including volume (over the entire domain) and surface areas (including individual inlet and outlet surfaces, such as the aorta, for which the total area was obtained by summing 12 different sections, as shown in Figure 1). These measurements allowed us to calculate the true theoretical values a priori, therefore defined "calculated"—for example, by calculating the average velocity at the inlet and outlet surface as the flow rate divided by the corresponding cross-sectional area.

The relative error for each quantity was then computed by comparing the values obtained post-processing from the software (derived from the CFD simulation and therefore called "simulated", reported in Section 3 as Derived Values) with the true theoretical values (calculated beforehand based on the geometry or values set such as the inlet flow rate and the outlet pressure).The relative error was therefore calculated as follows:

$$E_r = \frac{|\text{simulated value} - \text{calculated value}|}{\text{calculated value}} \tag{1}$$

Relative errors for the aorta and RCA are presented in Section 3.

3. Results

In this section, we present the simulation results obtained using COMSOL post-processing tools, focusing on the velocity field under the assumption of rigid vessel walls (no fluid–structure interaction). Several studies have investigated the impact of arterial geometry and treatment approaches on hemodynamics using computational fluid dynamics (CFD) simulations, highlighting the relevance of flow distribution and wall shear stress parameters [25–27].

3.1. Flow Behavior in the Aorta

We focused on the visualization of streamlines, as shown in Figure 3. In particular, it provides a detailed and intuitive representation of the velocity field intensity distribution along the yz-plane. The highest velocities are observed at the center of the epiaortic branches, with a maximum velocity of 2.7 m/s reached within the right carotid artery, originating from the brachiocephalic artery. Significant velocities are also present in the lower outlet branches, particularly in the renal and mesenteric arteries.

Figure 3 illustrates the streamlines of the flow considering the Newtonian model, confirming that the flow remains laminar under these physiological conditions.

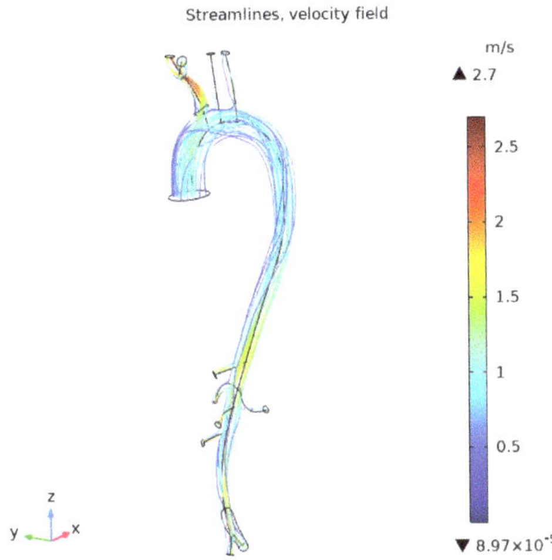

Figure 3. Aorta - Newtonian model: streamlines.

A cutting plane (shown in Figure 4a) was created near the inlet section, which also intersects the descending aorta just downstream of the aortic arch, and in these sections, velocity was observed (Figure 4b).

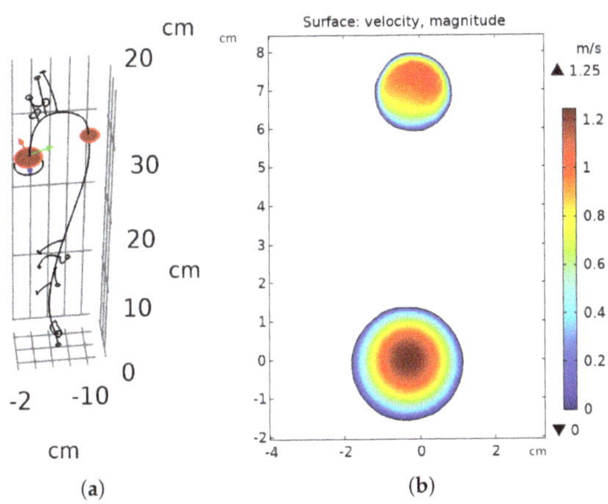

Figure 4. Aorta - Newtonian model: (**a**) Cut plane, inlet. (**b**) Velocity magnitude.

As shown in Figure 4b, the maximum velocity is observed at the center of the vessel, where the velocity gradient is minimal, resulting in low shear rate values. Conversely, near the vessel walls, the velocity rapidly decreases to zero, creating steep velocity gradients that correspond to higher shear rate values. This distribution is consistent with the typical laminar flow profile, where shear stress is predominantly concentrated in the peripheral regions of the lumen.

Observations are similar in the epiaortic branches, as shown in Figure 5, still considering the Newtonian model.

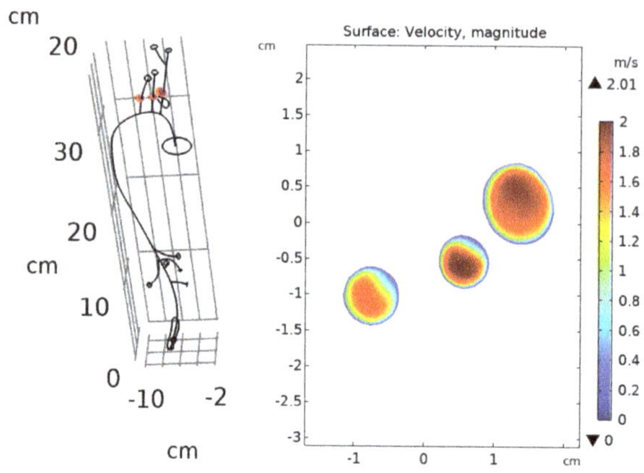

Figure 5. Aorta - Newtonian model: (**a**) Cut plane, epiaortic branches. (**b**) Velocity magnitude.

The same procedure was then followed for the other five models as well. In Figure 6, for example, we can see the velocities reached within the epiaortic vessels for each model. Compared to the Newtonian model, the shear-thinning models underestimate the maximum velocity, while the viscoelastic models overestimate it. The exact opposite occurs for the shear rate.

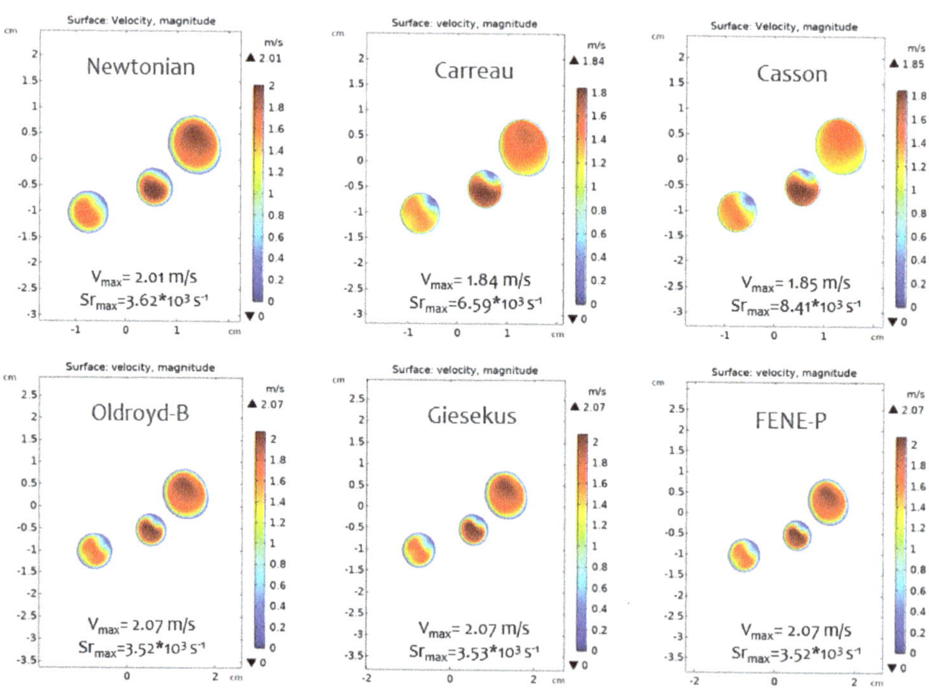

Figure 6. Aorta models - epiaortic branches.

All the post-processing values were derived from the software: for example, the average speeds (on the input and output surfaces), the flow rates, etc., which are reported in Table 1. Then, following the algorithm described in Section 2.5, the relative error was calculated for each model. Relative errors are reported in Table 2.

Table 1. Derived values.

Parameter	Value	Unit
vel (line average)	1.25080	m/s
volume-averaged velocity	0.84300	m/s
surface-averaged velocity (in)	0.61699	m/s
surface-averaged velocity (out)	1.45420	m/s
volume-averaged pressure	12,679.00	Pa
surface-averaged pressure (in)	13,600.00	Pa
surface-averaged pressure (out)	9372.50	Pa
Qin	4.50×10^{-4}	m^3/s
Qout	4.61×10^{-4}	m^3/s

Table 2. Relative error—AORTA.

Parameter	True Value	Newtonian	Carreau	Casson	Oldroyd-B	Giesekus	FENE-P
Average velocity on surface (in)	0.617 m/s	0.000	0.000	0.000	0.000	0.000	0.000
Average velocity on surface (out)	1.417 m/s	0.025	0.028	0.027	0.029	0.029	0.029
Average pressure on surface (out)	9332.54 Pa	0.004	0.002	0.002	0.002	0.002	0.002
Qin	4.50×10^{-4} m^3/s	0.000	0.000	0.000	0.000	0.000	0.000
Qout	4.50×10^{-4} m^3/s	0.025	0.028	0.026	0.042	0.029	0.042
Avg Relative Error [1]		**0.011**	**0.012**	**0.011**	**0.015**	**0.012**	**0.015**

[1] Highlighted in bold for each model, which serves as a key comparative indicator.

3.2. Flow Behavior in the RCA

The same analysis was performed on the coronary model. From the streamlines (Figure 7), related to the Newtonian model, it is evident that the highest velocities are reached at the point of maximum curvature and at the outlet. Note how the flow is laminar under these physiological conditions.

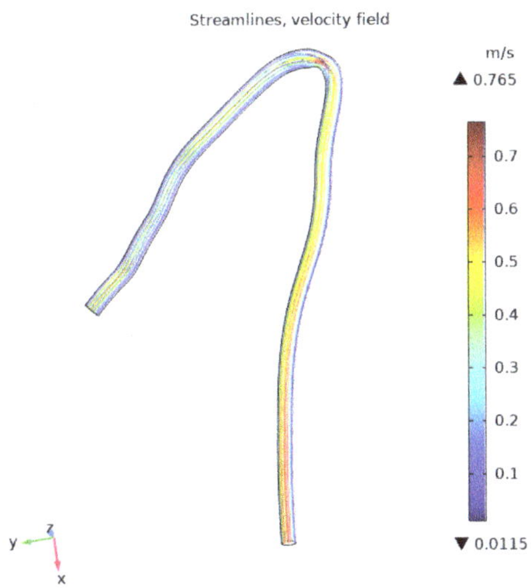

Figure 7. RCA - Newtonian model: streamlines.

The velocity magnitude was analyzed in greater detail at the section corresponding to the maximum curvature (Figure 8) and the outlet section (Figure 9).

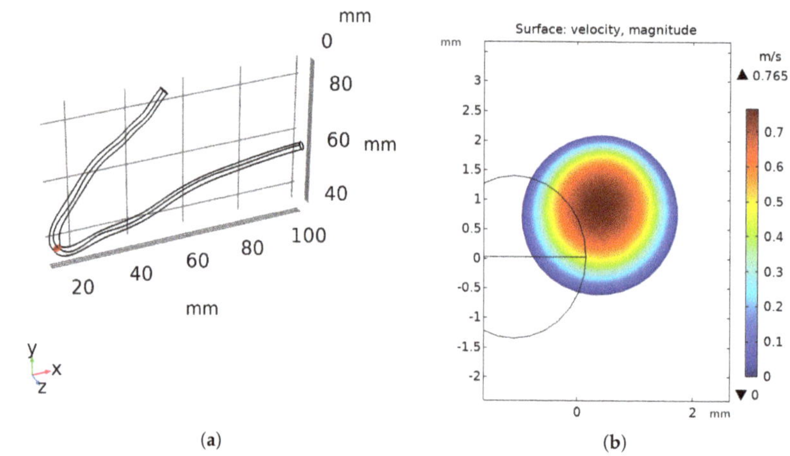

Figure 8. RCA - Newtonian model: (**a**) Cut plane, maximum curvature. (**b**) Velocity magnitude.

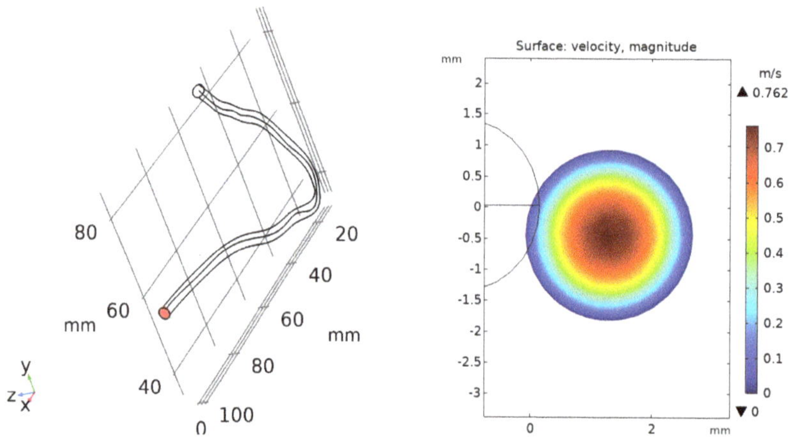

Figure 9. RCA - Newtonian model: (**a**) Cut plane, outlet. (**b**) Velocity magnitude.

The same procedure was followed for the other five models as well. In Figure 10, for example, we can see the velocities reached in the outlet section for each model. Compared to the Newtonian model, the shear-thinning models underestimate the maximum velocity, while the viscoelastic models overestimate it. The exact opposite occurs for the shear rate.

Then all the post-processing values were derived from the software and the relative error was calculated for each model. Relative errors are reported in Table 3.

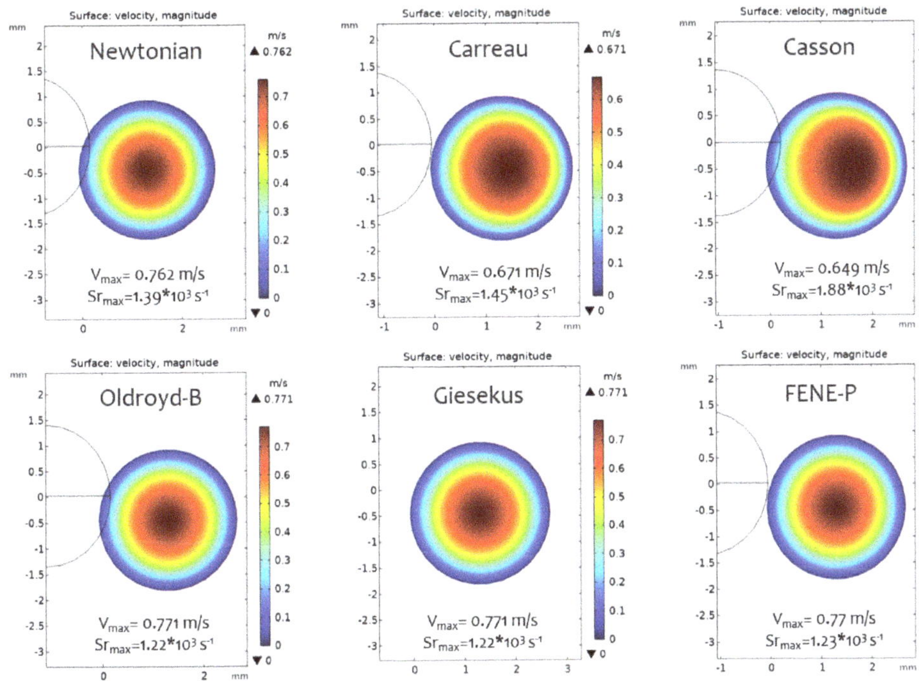

Figure 10. RCA models - outlet section.

Table 3. Relative error—RCA.

Parameter	True Value	Newtonian	Carreau	Casson	Oldroyd-B	Giesekus	FENE-P
Average velocity on volume	0.34300 m/s	0.118	0.110	0.109	0.110	0.110	0.108
Average velocity on surface (in)	0.29338 m/s	0.008	0.012	0.014	0.008	0.008	0.012
Average velocity on surface (out)	0.39355 m/s	0.023	0.028	0.035	0.022	0.022	0.021
Average pressure on surface (out)	17,731.83 Pa	0.001	0.000	0.000	0.000	0.000	0.000
Qin	2.25×10^{-6} m^3/s	0.000	0.000	0.000	0.000	0.000	0.000
Qout	2.25×10^{-6} m^3/s	0.025	0.000	0.000	0.006	0.006	0.007
Avg Relative Error [1]		**0.026**	**0.025**	**0.026**	**0.024**	**0.024**	**0.025**

[1] Highlighted in bold for each model, which serves as a key comparative indicator.

4. Discussion

The results of this study suggest that the Newtonian approximation is suitable for modeling flow in large-caliber vessels, such as the aorta. The Newtonian model showed the highest accuracy for aortic flow, with a relative error of 1.1%, and significantly shorter simulation times compared to more complex models. This confirms that the Newtonian approximation is adequate for large-caliber vessels. In contrast, while the Casson model produced a similar error to the Newtonian model, it required simulation times approximately five times longer.

Viscoelastic models did not show any advantage in simulations of aortic flow, where the flow is more stable and the viscoelastic behavior of blood has a lesser impact. However, for smaller caliber vessels, such as the right coronary artery (RCA), having more variable geometry and flexibility and being subjected to less stable flow conditions, viscoelastic models proved more accurate. This aligns with findings from earlier work [6], which showed that viscoelastic models are better suited for capturing complex fluid dynamics in smaller vessels.

For coronary flow, shear-thinning models represent a viable alternative, offering a good balance between computational efficiency and accuracy. These models provide valuable insights into the non-Newtonian nature of blood flow, supporting previous studies that have emphasized their utility in simulating real-world fluid dynamics. The shear-thinning behavior is particularly relevant in microvascular and pathological flow conditions, where blood exhibits significant changes in viscosity under varying shear rates.

From the perspective of our working hypothesis, the findings confirm that model selection depends not only on the vessel size but also on the specific characteristics of the blood flow. Our results are consistent with the hypothesis that more complex models, while computationally demanding, yield more accurate predictions for smaller vessels and pathological states.

The broader implications of these findings extend to clinical applications, where accurate flow models are crucial for simulating pathological conditions, such as atherosclerosis, and for planning personalized treatment strategies. Future research should focus on developing hybrid models that integrate the strengths of various approaches, such as combining Newtonian and non-Newtonian elements, to improve model accuracy while maintaining computational feasibility.

While numerical errors such as discretization and truncation are inherent to any finite element-based simulation, the observed variation in relative errors across rheological models is not random. On the contrary, it reflects the degree of alignment between the model's theoretical assumptions and the hemodynamic conditions of the specific vascular region. For instance, the superior performance of viscoelastic models in the RCA suggests that these formulations are better suited to capturing flow behaviors in small-caliber, tortuous vessels under non-uniform conditions. Therefore, beyond pure accuracy, the error distribution itself becomes a useful indicator of model suitability—offering a comparative understanding of when and where each rheological approach is most effective. This interpretation complements the working hypothesis and supports the idea that model selection should be context-driven, based on anatomical and physiological relevance as much as on numerical output.

As discussed in Section 2.3 regarding the boundary conditions, the current simulations assume rigid vessel walls, which simplifies the computational setup, making this approximation a reasonable compromise between accuracy and efficiency for current clinical applications. We reiterate that in some cases, particularly when vascular stiffness is markedly increased, this assumption may be entirely valid. Nevertheless, we acknowledge that arterial walls exhibit elastic and viscoelastic behavior, particularly in large vessels such as the aorta, and incorporating FSI models in future studies would enable a more physiologically accurate representation of hemodynamic behavior. Therefore, in this study, we adopted the rigid wall assumption and used COMSOL Multiphysics® to focus on isolating and comparing the effects of different rheological models on blood flow under controlled, physiologically realistic conditions. Although this approach does not include direct experimental validation, it allows for a systematic and reproducible comparison across models, which is a critical step in understanding their relative predictive behavior. The rigid wall approximation is widely accepted in preliminary CFD studies, particularly when vessel wall properties are uncertain or patient-specific data are unavailable. It also offers a favorable balance between computational efficiency and physiological realism, especially in scenarios where increased vascular stiffness justifies this simplification. Furthermore, although the graphical outputs generated by COMSOL Multiphysics® are primarily intended for visualization, they are grounded in numerically robust solutions of the governing equations and can provide meaningful qualitative and quantitative insight when interpreted in conjunction with detailed numerical data. Accordingly, the simulation framework employed—despite its inherent simplifications—represents a methodologically sound and computationally efficient platform for the systematic comparison of rheological models under physiologically motivated boundary conditions.

Finally, it should be noted that the model is steady-flow but arterial flows are not quasi-stationary—i.e., the flow field at an instant of a cardiac cycle is not the same as a steady flow at that instantaneous flow rate. The flow differs on acceleration and deceleration. This represents a limitation.

An additional limitation of this study is the lack of direct experimental validation (e.g., by Doppler ultrasound or MRI) for the specific vascular geometries analyzed. However, an indirect validation strategy was adopted, comparing the simulation results with theoretical values based on physiological data available in the literature. This approach, commonly used in similar CFD studies, provided good agreement and supports the reliability of the simulation results.

5. Conclusions

This study demonstrates the applicability and potential of computational fluid dynamics and distributed parameter modeling in providing insights into the hemodynamic environment of the human aorta and the right coronary artery. By integrating patient-specific geometries with physiologically consistent boundary conditions, we were able to estimate key indicators such as pressure drops, velocity profiles, and wall shear stress in a realistic manner. The most significant physical outcome of this work lies in the ability of these models to predict local flow behaviors that may be linked to the onset and progression of cardiovascular diseases, particularly in regions prone to disturbed flow. This capability not only reinforces the role of mathematical modeling in preclinical assessments but also supports its translational potential in personalized medicine. The agreement between model outputs and known physiological ranges supports the reliability of the simulations and encourages further validation with clinical data.

The study explores how different rheological models influence simulation outcomes under controlled and physiologically realistic conditions. The use of the "theoretical values"—derived from the imposed boundary conditions and literature-based parameters—was intended as a benchmark for internal consistency. So, the comparison across models serves as a relative performance evaluation under consistent computational settings. The authors do not claim to establish absolute accuracy without validation against experimental or gold-standard numerical data, hence the need for future studies to incorporate MRI-based or particle image velocimetry (PIV) datasets for proper model validation [28–30].

Author Contributions: Conceptualization, M.M., V.G. and G.F.; methodology, M.M., V.G., A.P. and G.F.; software, M.M., V.G., A.P. and G.F.; validation, G.F.; formal analysis, M.M., V.G. and G.F.; investigation, M.M., V.G. and G.F.; resources, G.F.; data curation, M.M., V.G. and G.F.; writing—original draft preparation, M.M., V.G. and G.F.; writing—review and editing, A.P. and M.M.; visualization, M.M., V.G. and G.F.; supervision, V.G. and G.F.; project administration, G.F.; funding acquisition, G.F. All authors have read and agreed to the published version of the manuscript.

Funding: This research received no external funding.

Institutional Review Board Statement: Not applicable.

Informed Consent Statement: Not applicable.

Data Availability Statement: The original contributions presented in the study are included in the article, further inquiries can be directed to the corresponding author.

Conflicts of Interest: The authors declare no conflicts of interest.

References

1. Kim, Y.; Park, J.; Kim, M. Diagnostic approaches for inherited hemolytic anemia in the genetic era. *Blood Res.* **2017**, *52*, 84–94. [CrossRef] [PubMed]
2. Formaggia, L.; Perktold, K.; Quarteroni, A. Basic mathematical models and motivations. In *Cardiovascular Mathematics: Modeling and Simulation of the Circulatory System*; Quarteroni, A., Formaggia, L., Veneziani, A., Eds.; Springer: Milan, Italy, 2009; pp. 47–75.
3. Piatti, F.; Belkacemi, D.; Caimi, A.; Sturla, F.; Greiser, A.; Pluchinotta, F.; Redaelli, A. On the potential of 4D Flow in guiding CFD analyses: A case study of aortic coarctation. In Proceedings of the VII Meeting Italian Chapter of the European Society of Biomechanics (ESB-ITA 2017), Rome, Italy, 28–29 September 2017; pp. 28–29.
4. Ramakrishnan, K.; Shailendhra, K. Hydromagnetic blood flow through a uniform channel with permeable walls covered by porous media of finite thickness. *J. Appl. Fluid Mech.* **2013**, *6*, 39–47.
5. Hoque, K.E.; Ferdows, M.; Sawall, S.; Tzirtzilakis, E.E.; Xenos, M.A. The impact of hemodynamic factors in a coronary main artery to detect the atherosclerotic severity: Single and multiple sequential stenosis cases. *Phys. Fluids* **2021**, *33*, 31903. [CrossRef]
6. Fernandes, M.; Sousa, L.C.; António, C.C.; Pinto, S.I.S. Accuracy and temporal analysis of non-Newtonian models of blood in the computational FFR—Numerical implementation. *Int. J. Non-Linear Mech.* **2024**, *161*, 104683. [CrossRef]
7. Rafiei, A.; Saidi, M. Numerical evaluation on blood rheological behavior in a realistic model of aneurysmal coronary artery. *Comput. Methods Biomech. Biomed. Eng.* **2021**, *6*, 579–596.
8. Yilmaz, F.; Gundogdu, M.Y. A Critical Review on Blood Flow in Large Arteries: Relevance to Blood Rheology, Viscosity Models, and Physiologic Conditions. *Korea-Aust. Rheol. J.* **2008**, *20*, 197–211.
9. Campo-Deaño, L.; Dullens, R.; Aarts, D.G.; Pinho, F.T.; Oliveira, M.S. Viscoelasticity of blood and viscoelastic blood analogues for use in polydimethylsiloxane in vitro models of the circulatory system. *Biomicrofluidics* **2013**, *7*, 034102. [CrossRef]
10. Fernandes, M.; Sousa, L.C.; António, C.A.C.; Pinto, S.I. Modeling the Five-Element Windkessel Model with simultaneous utilization of blood viscoelastic properties for FFR achievement: A proof-of-concept study. *Mathematics* **2023**, *11*, 4877. [CrossRef]
11. Comsol. COMSOL-Software for Multiphysics Simulation. Available online: https://www.comsol.com/ (accessed on 10 February 2025).
12. Caruso, M.V.; Renzulli, A.; Fragomeni, G. Influence of IABP-Induced Abdominal Occlusions on Aortic Hemodynamics: A Patient-Specific Computational Evaluation. *ASAIO J.* **2017**, *63*, 161–167. [CrossRef]
13. Caruso, M.V.; De Rosa, S.; Indolfi, C.; Fragomeni, G. Computational analysis of stenosis geometry effects on right coronary hemodynamics. In Proceedings of the Engineering in Medicine and Biology Society (EMBC), 2015 37th Annual International Conference of the IEEE, Milan, Italy, 25–29 August 2015; pp. 981–984.
14. Belkacemi, D.; Al-Rawi, M.; Abbes, M.T.; Laribi, B. Flow behaviour and wall shear stress derivatives in abdominal aortic aneurysm models: A detailed CFD analysis into asymmetry effect. *CFD Lett.* **2022**, *14*, 60–74. [CrossRef]
15. Caruso, M.V.; Gramigna, V.; Fragomeni, G. A CFD investigation of intra-aortic balloon pump assist ratio effects on aortic hemodynamics. *Biocybern. Biomed. Eng.* **2019**, *39*, 224–233. [CrossRef]
16. Perinajová, R.; Juffermans, J.F.; Westenberg, J.J.M.; van der Palen, R.L.; van den Boogaard, P.J.; Lamb, H.J.; Kenjereš, S. Geometrically induced wall shear stress variability in CFD-MRI coupled simulations of blood flow in the thoracic aortas. *Comput. Biol. Med.* **2021**, *133*, 104385. [CrossRef]
17. Friedman, M.H.; Bargeron, C.B.; Duncan, D.D. Effects of arterial compliance and non-Newtonian rheology on correlations between intimal thickness and wall shear. *J. Biomech. Eng.* **1992**, *114*, 317. [CrossRef]
18. Brown, A.G.; Shi, Y.; Marzo, A.; Staicu, C.; Valverde, I.; Beerbaum, P.; Lawford, P.V.; Hose, D.R. Accuracy vs computational time: Translating aortic simulations to the clinic. *J. Biomech.* **2012**, *45*, 516–523. [CrossRef]
19. Suo, J.; Oshinski, J.; Giddens, D.P. Effects of wall motion and compliance on flow patterns in the ascending aorta. *J. Biomech. Eng.* **2003**, *125*, 347–354.
20. Romarowski, R.M.; Lefieux, A.; Morganti, S.; Veneziani, A.; Auricchio, F. Patient-specific CFD modelling in the thoracic aorta with PC-MRI-based boundary conditions: A least-square three-element Windkessel approach. *Int. J. Numer. Methods Biomed. Eng.* **2018**, *34*, e3134. [CrossRef]
21. Qiao, Y.; Mao, L.; Ding, Y.; Zhu, T.; Luo, K.; Fan, J. Fluid-structure interaction: Insights into biomechanical implications of endograft after thoracic endovascular aortic repair. *Comput. Biol. Med.* **2021**, *138*, 104882. [CrossRef]
22. Xi, Y.; Li, Y.; Wang, H.; Sun, A.; Deng, X.; Chen, Z.; Fan, Y. The impact of ECMO lower limb cannulation on the aortic flow features under differential blood perfusion conditions. *Med. Nov. Technol. Devices* **2022**, *16*, 100183. [CrossRef]
23. Hsiao, H.-M.; Lee, K.H.; Liao, Y.C.; Cheng, Y.C. Hemodynamic simulation of Intra-stent Blood Flow. *Procedia Eng.* **2012**, *36*, 128–136. [CrossRef]
24. Hoque, K.E.; Wu, P.; Li, Z.; Zhao, X. The effect of hemodynamic parameters in patient-based coronary artery models with serial stenoses: Normal and hypertension cases. *Comput. Methods Biomech. Biomed. Eng.* **2020**, *23*, 1001–1008. [CrossRef]

25. Gramigna, V.; Caruso, M.V.; Fragomeni, G. A computational fluid dynamics study to compare two types of arterial cannulae for cardiopulmonary bypass. *Fluids* **2023**, *8*, 302. [CrossRef]
26. Wen, J.; Zhang, X.; Li, Y.; Sun, Q. A computational simulation of the effect of hybrid treatment for thoracoabdominal aortic aneurysm on the hemodynamics of the abdominal aorta. *Sci. Rep.* **2016**, *6*, 23801. [CrossRef] [PubMed]
27. Buradi, A.; Mahalingam, A. Effect of stenosis severity on wall shear stress-based hemodynamic descriptors using multiphase mixture theory. *J. Appl. Fluid Mech.* **2018**, *11*, 1497–1509. [CrossRef]
28. Miyazaki, S.; Itatani, K.; Furusawa, T.; Nishino, T.; Sugiyama, M.; Takehara, Y.; Yasukochi, S. Validation of numerical simulation methods in aortic arch using 4D Flow MRI. *Heart Vessels.* **2017**, *32*, 1032–1044. [CrossRef]
29. Xie, X.; Li, N.; Yu, T.; Pu, W.; Lin, H.; Li, X.; Li, G.; Li, C.; Chen, Y.; Lian, K. Validation of a Novel Computational Fluid Dynamics Based Method for Assessing Intracoronary Flow: Combining Coronary Angiography and Fractional Flow Reserve. In *Catheterization and Cardiovascular Interventions: Official Journal of the Society for Cardiac Angiography & Interventions*; Wiley: New York, NY, USA, 2025.
30. Caridi, G.C.A.; Torta, E.; Mazzi, V.; Chiastra, C.; Audenino, A.L.; Morbiducci, U.; Gallo, D. Smartphone-based particle image velocimetry for cardiovascular flows applications: A focus on coronary arteries. *Front. Bioeng. Biotechnol.* **2022**, *10*, 1011806. [CrossRef]

MDPI AG
Grosspeteranlage 5
4052 Basel
Switzerland
Tel.: +41 61 683 77 34

Applied Sciences Editorial Office
E-mail: applsci@mdpi.com
www.mdpi.com/journal/applsci

www.ingramcontent.com/pod-product-compliance
Lightning Source LLC
Chambersburg PA
CBHW061626210926
51850CB00009B/2446